W9-CCH-710

SIMON & SCHUSTER

•

New York

London

Toronto

Sydney

Tokyo

Singapore

THE SCIENCE OF DESIRE

The Search for the Gay Gene and the Biology of Behavior

•

DEAN HAMER

AND PETER COPELAND

SIMON & SCHUSTER
Rockefeller Center
1230 Avenue of the Americas
New York, New York 10020

Designed by Liney Li
Manufactured in the United States of America

1 3 5 7 9 8 6 4 2

Library of Congress Cataloging-in-Publication Data
Hamer, Dean H.
The science of desire : the search for the gay gene and the
biology of behavior / Dean Hamer and Peter Copeland.
p. cm.
Includes bibliographical references and index.
1. Homosexuality—Genetic aspects.
2. Behavior genetics. 3. Sex-linkage (Genetics)
4. Sociobiology. 5. Sexual orientation.
I. Copeland, Peter. II. Title.
HQ76.25.H34 1994
304.5—dc20 94-22260 CIP
ISBN 0-671-88724-6

The names and certain identifying details of
research participants and other persons portrayed in this book
have been changed.

ACKNOWLEDGMENTS

I am grateful to the many people who contributed to this book and the research it describes. The most important were the men and women who volunteered for our study. I thank each and every one of them for their time, interest, and cooperation. I am also grateful to the four outstanding and dedicated scientists who collaborated on the linkage project: Stella Hu, Victoria Magnuson, Nan Hu, and Angela Pattatucci.

I could not have started a whole new line of research without the suggestions and help of many talented scientists. I especially thank Roger Gorski, Jin Zeng, Bennet Prickrill, William Gahl, Lawrence Charnas, Michael Pollis, Wesley McBride, Chris Amos, Elliot Gershon, Lynn Goldin, Jeremy Nathans, David Goldman, Eric Lander, Michael Boehnke, Francis Collins, Juanita Eldridge, Simon LeVay, Michael Bailey, and Richard Pillard. I also thank the members of our Advisory Committee: Fred Bonkovsky, Sandy Chamblee, Peter Hawley, Jeanne Mackenzie, James Weinrich, Alison Wichman, Pepper Schwartz, and Tom Sauerman. Our research would not have been possible without the assistance and cooperation of the staff members at the NIH Interinstitute Genetics Clinic, the NIAID HIV Clinic, the NIAAA alcoholism program, and the Whitman-Walker Clinic.

I owe a special thanks to James Weinrich, who was involved in the early stages of planning this book and provided me with unique guidance in both the theory and practice of sexuality research. Dr. Weinrich, and also Robert Trivers, were especially important to my thinking about the evolution of sexual orientation.

For their comments and suggestions on the manuscript, I thank my editor Bob Bender, my agent Julie Castiglia, and my colleagues Elliot Gershon, Mike Bailey, and Roger Gorski.

Last but not least, I thank the National Institutes of Health and the National Cancer Institute for supporting my research for the past seventeen years.

Although there are two authors of this book, it is written in the first person because I, Dean Hamer, am solely responsible for its scientific content and interpretation. My cowriter and friend Peter Copeland is a journalist who undertook the often onerous task of translating my thoughts into words that we hope both scientists and nonscientists will be able to understand and enjoy.

In memory of

Marilou Hielman Hamer,

a very special mother

CONTENTS

WE DO NOT EVEN IN THE LEAST KNOW

THE FINAL CAUSE OF SEXUALITY.

THE WHOLE SUBJECT IS HIDDEN IN DARKNESS.

—Charles Darwin, 1862

PREFACE

Sex is important. It's important in biology because sex is the source of continued life and the mechanism by which higher organisms pass on their genes and evolve. It's important to health because so many diseases, of which only the most devastating example is AIDS, are sexually transmitted. And it's important to humanity because it is the source of so much of our joy, frustration, pleasure, anguish, pursuit, and thought.

Given the significance of sexuality, one might expect it to be the subject of a large amount of research. It is not. There are no federally supported sexuality research centers. Scientists and academics who try to make a career out of analyzing sexuality find their way blocked by funding sources and tenure committees. The scientific literature contains more articles on the genetics of eye color in fruit flies than on the biology of human sexuality. And we spend far more money treating the results of sexually transmitted diseases than trying to prevent them.

Why is sexuality so understudied and misunderstood? Some say that sex is too private and intimate to be openly analyzed and discussed. I disagree. A topic that impinges on the very existence of our species ought to be studied under the brightest light available.

Others say that sexuality is a moral issue, better left to the church than to science. I disagree again. The clergy is no better equipped now to understand the origins of sexuality than it was a century ago to comprehend the origins of life or four centuries ago to fathom the nature of the solar system.

And then there are those, scientists among them, who say that sexuality is too mysterious and complex to ever be analyzed or comprehended. I disagree with them, too. To accept this attitude would be to negate the tremendous advances that genetics, neurobiology, evolution, and psychology have made in our understanding of the human mind. To give up on understanding sex is to surrender to ignorance, to despair of our own potential for thought and knowledge.

The aim of this book is to show that human sexuality can, and indeed must, be scientifically studied. What do I mean by the "scientific" study of sex? First, the study must be based on carefully controlled observations and experiments, not on the hearsay, innuendo, and myths that characterize so much of the discourse on human sexuality and behavior. Second, it must produce specific and testable predictions, not just vague generalities that defy empirical examination. And lastly, it must ultimately be based on physical laws rather than on appeals to "nature's way" or "God's will" or the like.

These are the principles that guided my search for the "gay gene" and led to the discovery that is the main topic of this book: the finding of a linkage between male homosexuality and DNA markers on the X chromosome. The first two thirds of this book describe some of the trials and tribulations that led to this finding, together with the scientific basis for the research design. The final third of the book speculates on how "gay genes" might work, how they might have evolved, and their implications for the origins of other human behaviors from aggressiveness to shyness. The final chapter discusses my views on the social, legal, and ethical ramifications of research on sexuality, a matter of continuing—and healthy—debate.

My purpose in writing this book is not to offer a complete or polished theory of sexual orientation but rather to describe what we found using the tools of modern genetics, how biological findings can broaden rather than narrow our understanding of the diversity of human sexual expression, and what remains to be learned. Along the way, I hope to debunk some of the more common myths about the role of biology in human sexuality and behavior, especially as concerns the question of homosexuality, and to give a sense of how scientific research is actually performed. If this book can make even a small contribution to the scientific understanding of sex, I shall be satisfied with my efforts.

Chapter One

THE SEARCH FOR A "GAY GENE"

Every single day, in laboratories and universities, scientists make discoveries, some great and some small, but few of them are reported on the front pages of the world's major newspapers, featured on talk shows, included in *Time* magazine's list of "The Best Science" of the year, heralded in the *National Enquirer,* and turned into catchy slogans for T-shirts.

Once in a while, though, a study will hit the right buttons at the right time and will cause a wave of scientific and popular reaction that, for the scientists involved, can be both encouraging and frightening. That's what happened to me in the summer of 1993, when a scientific journal reported the results of a two-year study performed in my normally quiet U.S. government laboratory. The topic was a genetic link to homosexuality, and the study appeared when gay issues were at the top of the political agenda because of the rancorous debate over homosexuals in the military.

The day the study appeared, the front-page *Washington Post* story bore the straightforward headline: STUDY LINKS GENES TO HOMOSEXUALITY. *USA Today*'s headline asked, IS THERE A GAY GENE? That evening I was invited to appear on "Nightline" for the first of many television interviews.

The follow-up stories began to raise the implications of our study. GAYS AND THE DNA LINK: STUDY SHOWING THAT HOMOSEXUALITY MAY RUN IN THE FAMILY SETS OFF ALARM BELLS, warned Canada's *Edmonton Journal.* Another Canadian paper, the *Ottawa Citizen,* wrote, GENE FIND OPENS PANDORA'S BOX OF ETHICAL AND LEGAL ISSUES.

Some headlines were alarmist, like the *Daily Telegraph* of London's darkly foreboding CLAIM THAT HOMOSEXUALITY IS INHERITED PROMPTS FEARS THAT SCIENCE COULD BE USED TO ERADICATE IT. Others apparently were meant to be reassuring: STUDY LINKING GENES TO HOMOSEXUALITY DOESN'T ALTER RELIGIOUS LEADERS' STANCE (Louisville, Kentucky, *Courier-Journal*). Many papers welcomed the study. As North Carolina's *Greensboro News & Record* editorialized, "Research might lead to more tolerance of gays."

Then skeptics appeared in the form of guest columnists, such as the medical doctor writing in the *San Francisco Examiner* who advised that reports of a genetic link to homosexuality smelled like a "fish story." The *National Enquirer* sounded a more "positive" note with the promise: SIMPLE INJECTION WILL LET GAY MEN TURN STRAIGHT, DOCTORS REPORT.

Many of the reactions were more personal, and my mailbox filled with letters from people thanking me for doing the study. Other letters promised I would burn in hell. One woman wrote to express her "amazement and disgust," while a "gay man and a molecular geneticist" writing from England castigated me for results that he felt would increase hatred of gays. "I fear that your work will make our lives more unbearable," he wrote. "Maybe this is your aim?" A man from the Virgin Islands, who accused me of "attempting to gain publicity and legitimize a purely moral question through science," ended his angry missive with "Take that, you scientist."

More common, and often very moving, were letters from people writing about their own families. One man explained that he was the father of two young men who recently had announced, to the surprise and dismay of the rest of the family, that they were gay. Badly shaken by the news, the father sought help from his church. He was told that homosexuality was wrong but that it could be cured through prayer. Most damning of all, he was told that it was caused by parents who lacked faith or didn't properly raise their children. The burden of guilt was tremendous, he wrote, and his first reaction was to hate himself, reject his sons, and deny what they had told him.

Then he saw an article about my work that said homosexuality might be genetic. His sense of relief was overwhelming because *he* wasn't the problem, after all. This was something out of his control; it was nature, not his nurture. He could forgive himself, and more importantly, forgive his sons and welcome them back into the loving embrace of the family, now made stronger by this test of faith.

Perhaps I should have been gratified by testimony like this. Every scientist wants to think his or her work matters, that long hours of tedious research won't just be filler for some little specialized journal but will improve lives and make the world a better place.

Instead, I was saddened. This man had changed the course of his life, and the lives of everyone in his family, because of a few paragraphs in a magazine. He decided to forgive his sons because we found a genetic link to homosexuality. But what if the experiment had failed? Or what if we gave his family a blood test and found they didn't have the "gay gene," that the sons were gay for some other reason? Then would this father go back to blaming himself for raising two gay sons, and would they be less worthy of his love?

These kinds of questions, and all the attention, were new to me, an obscure molecular geneticist who had spent the previous sixteen years bustling about in a small federal laboratory, jumbled floor to ceiling with beakers, vials, and other paraphernalia. It is the kind of place that has "emergency showers" in case something spills, but on most days not much happens that is very exciting. No one outside of my immediate scientific circle ever had paid much attention to my work before, but now strangers were sending me letters, reporters wanted interviews, lawyers subpoenaed me to testify in court, and members of Congress wanted to know just what in the world was going on.

All because I wondered what makes people gay.

The origins of human sexuality, and of homosexuality in particular, have puzzled philosophers, theologians, and ordinary people for thousands of years. In a few scattered cultures, homosexuality has been regarded as a normal part of life or even as a special talent or gift from the gods. In most societies we know, however, same-sex attractions or homosexual behavior have been treated as an unforgivable sin or a terrible crime. Beginning in the late 1800s, psychiatrists and psychologists turned their attention to homosexuality and concluded that it was a mental disorder caused by a misguided upbringing. This disease model remained the primary way of thinking about homosexuality during most of the twentieth century.

More recently, however, some scientists have begun to view both heterosexuality and homosexuality as natural variations of the human condition that are at least as deeply rooted in nature as in nurture. During the past several years, researchers have detected minute but significant differences between the brains of heterosexual and homosexual men. Others have shown that genetically identical twins are more likely to both be gay than are brothers who aren't twins. These results suggest that homosexuality is at least partly inherited—a finding of no great surprise to gay men, most of whom feel they were "born that way."

No one had ever been able to *prove* that homosexuality was swayed by genes, however, until our study offered the most convincing evidence to date that sexual orientation was genetically influenced. That's why we were in magazines and on TV, and why I was receiving kindness from strangers. We didn't invent a new idea, we just showed it was true.

The first part of our study was something that could have been done years ago, only nobody bothered. We simply traced back the lineages of gay men, looking for signs of homosexuality in all the twigs and branches of their family trees. We drew orchards of these trees, going back as far as anyone could remember and stretching as wide as possible to include second cousins and great uncles. We found far more gays on the mother's side of the families than on the father's side, a pattern consistent with a special type of inheritance called sex linkage. The expression comes from the fact that the relevant gene is on one of the two sex chromosomes, in this case the X chromosome.

The second part of the study was something that never had or could have been done before because the scientific tools and techniques were brand-new. We looked directly at gay men's genetic information—their DNA, the long threadlike molecule that contains both the blueprints of life and the instructions for carrying them out. Using an approach called DNA linkage analysis, we found that a small region of the X chromosome, Xq28, appeared to be the same in an unexpectedly high proportion of gay brothers. This finding provided the first concrete evidence that "gay genes" really do exist and narrowed the location of one of them to a few million out of the several billion bits of information that make us human. What we found was a marker, a strip of DNA usually transmitted whole, rather than the smaller bit of DNA regarded as a single gene.

Our results were published on 16 July 1993, in *Science,* the technical journal of the American Association for the Advancement of Science. The title of our six-page article was not likely to win any awards for headline writing: "A Linkage Between DNA Markers on the X Chromosome and Male Sexual Orientation."* A pithier summation of the study appeared a few days later on T-shirts sold at gay and lesbian bookstores: "Xq28—Thanks for the genes, Mom!"

WHY ALL THE FUSS?

Most of the time, proving something that people already suspected doesn't cause such a stir, but sexual orientation is no ordinary topic. It's at the center of a fierce debate involving politics, the law, religion, ethics, and the origins and meaning of human behavior. Remember how etiquette books used to advise that certain topics were not appropriate for dinner-table conversation? Put a bunch of those topics together and you have the "gay gene" story.

Our results were published during the midst of the great debate over gays in the military. In fact, the date set by President Clinton for the Pentagon to have a new policy on homosexuality was 15 July 1993, just one day before our paper appeared. When the study was published, my phone rang off the hook with questions about how the

* The research article is reprinted in Appendix A.

results might affect the new Pentagon policy. Some people said our study proved that keeping gay men and lesbians out of the military was the same as the earlier discrimination against African-Americans. Others, however, including a guest who appeared with me on the program "Nightline," thought it now would be necessary to treat gay and lesbian service members as a "third sex" and segregate them in separate barracks.

Our research also had legal implications outside the military. The Supreme Court has made subtle but important distinctions between characteristics that are "immutable," or deeply ingrained and not easily changed, and those that are freely chosen. Many legal experts felt the evidence for a genetic link to homosexuality would strengthen the evidence for immutability and therefore cause tighter scrutiny of laws that permitted discrimination against gays and lesbians in housing, employment, or participation in the political process. Others, though, argued that immutability was a red herring and that the real issue was equal protection, not biology.

There were ethical, medical, and economic issues as well. Although our research did not provide any sort of test for the still hypothetical "gay gene," there were legitimate concerns that we were headed in that direction. If such a test were developed, might parents decide to screen the fetus for homosexuality, just as they do now for Down syndrome and other genetic defects? Would some doctors regard homosexuality as a genetic defect that should be "cured" or weeded out of the population? Would insurance companies charge men with the "gay gene" more for coverage or refuse to serve them, knowing the high risk of AIDS faced by gay men? These were possibilities that deeply worried many people, including myself. Making sure the results of our study were used ethically and responsibly would be at least as difficult as conducting the research itself.

Last but not least, there were implications for the hoary question of "nature versus nurture." Most scientists now agree that the very wording of this question represents a false dichotomy and that both biology and the environment play some role in virtually all human behaviors. The relative contributions and precise nature of these factors, however, remain a matter of considerable speculation and debate, largely because of the lack of tools to dissect them. Our finding of a molecular linkage for an obviously complex and diverse aspect of behavior, sexual orientation, suggests the same approach could be used to identify genes for many different facets of human existence.

The goal of scientists around the world is to know, sometime during the next five to fifteen years, the precise structure of every single one of the 100,000 or so genes that make up our inherited information. If we were able to find a gene for such a complex behavior as homosexuality knowing only a fraction of this information, won't others soon find genetic links for anger, impatience, or joy? Could there be genes that predispose some people to become great musicians and others to become criminals? And if such genes were found, wouldn't parents be tempted to manipulate or select the ones for characteristics they consider desirable? Everyone accepts the idea that genes give some people blue eyes and others brown eyes or make some people tall and others short. No one before our study, however, had proved a genetic linkage to a complex behavior—any kind of behavior, not just sexual behavior. Finding and proving more of these genetic links will change the way we understand ourselves and perhaps change the very future of the human race.

AT LAST, NOT YEAST

As a molecular geneticist at the National Institutes of Health (NIH), the world's largest biomedical research facility, I was in a perfect place to look for genes involved in sexuality and other aspects of human behavior. The NIH has more than 16,000 employees, including 3,700 doctoral-level researchers, 1,800 clinicians, and 7,800 professional and technical staffers spread out over a "campus" in Bethesda, Maryland, just outside Washington, D.C. There are 370 beds dedicated to research, and the most modern laboratories anywhere. The complex has specialized institutes dealing with a growing range of health issues from cancer to alcoholism to child development and mental health. A new genetic research institute, the Human Genome Center, opened during 1993.

The ideal place to do research on homosexuality would be the

National Institute of Sexuality. If only we had one. Despite the importance of sexuality to health—after all, sex is the very source of continued life—there is no federally funded research center for sexuality. In fact, when I started our research in 1991 there was not even a single research group or laboratory at the NIH devoted to sexuality.

The Cancer Institute, where I work, certainly is not noted for its research into human sexuality. My own work there was not about sex, either. In fact, my work didn't even involve human beings. My usual subjects were yeast cells and mice, which are far less complicated than people, and I was studying things about them that are far less mystifying than sexuality. For ten years my entire laboratory had worked on one problem: the regulation of metallothionein (MT) gene transcription by heavy metal ions. Our goal was to understand how genes are turned on and off, or regulated.

Briefly, metallothionein is a protein that binds to heavy metal ions, such as copper, cadmium, mercury, and zinc, to protect cells against poisoning. When there are no metals around, the cell doesn't make MT, but when metals are present, the cell does. No one knew how this worked until 1988 when we discovered the secret of MT gene regulation, at least in yeast cells. We found that metals bind to a regulatory protein that changes shape. The altered protein binds to DNA sequences close to the beginning of the MT gene, which activates it. Other proteins bind to the regulatory protein, forming a complex that turns the MT gene DNA into messenger RNA and finally into the protein. This is a neat example of a molecular switch and provides a simple model of the type of switch that might be involved in more fundamental life processes, such as cellular determination and differentiation.

This was pure, basic science with few obvious applications. Fortunately, the NIH recognizes the importance of basic science, and not every project that is funded has to have an immediate application to medicine. Though my own lab is part of the Cancer Institute, during ten years the only possible connection to cancer we had come up with involved the use of a cancer drug called cis-platinum. One problem with this drug is that after extended therapy, patients build up a resistance, partly because the cancer cells start making more MT, which inactivates the platinum in the drug. If we could understand how this process worked, we might be able to improve the treatment.

People often ask why I switched from a field as obscure as metallothionein research to one as controversial as homosexuality. The answer is the same that most scientists give for why they do what they do: a combination of curiosity, altruism, and ambition (especially curiosity, both personal and scientific), combined with one more factor—boredom. After twenty years of doing science, I had learned quite a bit about how genes work in individual cells, but I knew little about what makes people tick.

The turning point in my career occurred in 1991, during a pair of conferences I attended in Oxford, England. Our work on MT was recognized internationally, and I was invited to be a guest speaker at both meetings. I was honored, but when I opened the floor to discussion, I had an extreme case of déjà vu: the questions people asked were exactly the same ones that had been asked the very first time I had talked about MT ten years earlier. Another speaker talked about work at his lab that was farther along than ours, but instead of being inspirational, that news only depressed me more. I realized that even if I stuck with this research for another ten years, the best I could hope for was to build a detailed three-dimensional replica of our little regulatory model. It didn't seem like much of a lifetime goal.

So even though the meetings focused on our work and I was an invited speaker, my mind was elsewhere. The evidence for this comes from a simple test that applies to all scientists. To determine what scientists really are interested in, don't focus on what they are doing. That just shows what they have been funded for and what experiments they think will work and produce another publication. Focus instead on what they are reading. By this time I had almost stopped looking at my longtime favorite journals, *Molecular and Cellular Biology* and *Cell,* which have terrific reads like "Protein Translocation into Proteoliposomes Reconstituted from Purified Components of the Endoplasmic Reticulum Membrane." Instead, I carried in my traveling bag *Nature Genetics,* the *American Journal of Human Genetics,* and the *Journal of AIDS Research.*

What finally made up my mind to switch fields were two books that I picked up at a bookstore in Oxford during a lull between the conferences. The first was *Descent of Man, and Selection in Relation to Sex,* by Charles Darwin, published in 1871. I confess that even though Darwin is considered the father of modern biology, I never actually had read any of his work, except for the usual summaries in textbooks. I was surprised to find that fully three quarters of this 872-page book are devoted to "sexual selection," the process whereby natural selection favors certain traits that make either males or females more successful in mating and therefore passing on their genes. The result is a sexual dimorphism, which refers to the differences between the two genders in nearly every species, from peacocks to human beings. Darwin also paid a great deal of attention to behavior, not just physical characteristics, and seemed quite certain that variations in behavior between individuals and between species must be at least partly inherited. The implication is that even sexuality most likely has a significant genetic component.

The second book was *Not in Our Genes,* by Richard Lewontin and colleagues. I knew Lewontin, a professor at Harvard University, was a very respected scientist with a solid reputation in evolutionary and population genetics. I also knew that he had criticized the idea that behavior is genetic, arguing instead that it is a product of class-based social structures. Sometimes critics give the most useful advice on how experiments ought to be done, and I thought Lewontin might show me what needed to be explored, but from the first chapter's "dialectical" critique of the New Right's view of human nature, it was clear this was a political rather than a scientific book. And it wasn't even good politics.

What I did learn from Lewontin's book and others I read was that the genetics of behavior, and sexuality in particular, is an emotionally and politically charged topic. I was pretty sure that if I had attended a scientific meeting about sexual orientation instead of gene regulation I would not have been bored. I also realized that switching my research topic was going to take more than just a change in reading habits; I also would need a certain amount of bureaucratic skill—and not a little luck.

I wasn't thinking about any of that flying home from Oxford. Instead, I was fascinated to learn that Darwin seemed so convinced that behavior was partially inherited, even though when he was writing genes had not been discovered, let alone DNA. On the other hand, why was Lewontin, a formidable geneticist, so determined not to believe that behavior could be inherited? He couldn't disprove the genetics of behavior in a lab, so he wrote a political polemic against it. Maybe there was room for some real science here.

SEX ED

When I returned home from Oxford, the first thing I needed to do was to learn about sex, scientifically speaking. I set aside six months to give myself a crash course in sexology, human genetics, and sexually-related health matters. To start, I sat down at my computer and called up Medline, a computerized database of scientific articles and books collected by the National Library of Medicine, which is located on the NIH campus. First I typed in the key word "homosexuality" and the appropriate cross-references for genetics and searched the database back to 1966. I waited a few seconds while the computer hunted for the articles that contained those two topics. I planned to print out a list of references and then go through them one by one and call up the articles. When the computer finally beeped to indicate the search was completed, it had found only fourteen articles. To appreciate how few that is, consider that a search for "metallothionein" and "genes" pulled up 654 references.

Fewer than half the references to genetics and homosexuality were actual research papers based on experiments. The rest were reviews, rehashes of old data, speculation, and polemics. Sitting at my computer, I thought to myself, the bad news is not much is known about the genetics of sexual orientation. The good news is it won't take long to master the literature.

A substantial amount of the work in the field had been done by just two researchers: Richard Pillard, a professor of psychiatry at Boston University, and J. Michael Bailey, an assistant professor of psychology at Northwestern University in Evanston, Illinois. After reading their published papers, I called both men, told them what I was thinking about doing, and asked where their work was headed. Pillard said he was running a psychiatric hospital in Boston and had little time for research anymore, but he did offer to give me tips on recruiting subjects, conducting interviews, and preparing questionnaires. Mike Bailey said he was busy doing research on sexual orientation and promised to share with me some very interesting new results from a study of twins.

Bailey was coming to town for a meeting of the American Psychological Association, so I invited him to my house on a Saturday morning. A young researcher from a working-class background, Bailey worked for a few years as a high school math teacher before going back to school for his Ph.D. in clinical psychology. He impressed me as a serious scientist in that he was more interested in data than in rhetoric. We share the Baconian concept that scientific models should be based on observation rather than on what a scientist already is convinced is true.

Bailey's study of twins was exciting. Until the development of DNA-based methods, studies of twins provided virtually the only tool to tease apart the roles of inheritance and environment in behavior. The basic idea is simple. Identical (monozygotic) twins share 100 percent of their genetic information, but fraternal (dizygotic) twins share only 50 percent of their genes. Therefore, if a trait is genetically influenced, identical twins should share the trait more often than fraternal twins, who in turn should display more similarity than unrelated people. By contrast, if all three categories of individuals display similar rates of the trait, then genes probably are not playing a major role.

Bailey used advertisements in gay newspapers to recruit 110 pairs of twins. He knew one twin in each pair was gay and determined the sexual orientation of the co-twin by sending him a questionnaire, or when that was not possible, by asking the brother. The results were striking. Out of 56 pairs of identical twins, 52 percent of the co-twins also were gay. But out of 54 pairs of fraternal twins, only 22 percent of the co-twins were gay. In other words, identical twins were more similar than fraternal twins, who in turn were more likely to be gay than the population at large. This was exactly what would have been expected for a characteristic that was influenced, albeit not determined, by genes. From his data, Bailey calculated that close to 50 percent of male sexual orientation could be linked to genetics. He wasn't saying that genes directly determined sexual orientation but that in the people he studied genes appeared to account for at least 30 percent and possibly as much as 70 percent of the reason they were gay or straight.

Bailey's results, which had not yet been published, had a major impact on my thinking. Although several previous reports on sexual orientation in twins had been published, all of them were blemished by inappropriate methodology or small sample sizes. Bailey's work was carefully performed and analyzed by an investigator with a solid background in behavioral genetics and statistics. The simplest interpretation of the results was that genes play a substantial role in determining male homosexuality. Moreover, the almost perfect 2:1 ratio of gay identical twins to fraternal twins suggested the genetic mechanism might not even be that complicated.

There was, however, one glaring anomaly in Bailey's data. Some of the brothers he looked at were adopted and not genetically related. Eleven percent of the adopted brothers were gay, compared with only 9.2 percent of the blood brothers who weren't twins. That meant genetically unrelated brothers were more likely to be gay than brothers who shared 50 percent of their inherited information, which made no sense genetically. According to the hypothesis, ordinary brothers and fraternal twins should have had the same chance of being gay, and the rate should have been higher than in the genetically unrelated, adopted brothers. Bailey speculated that the discrepancy might have been caused by the different ways the twins and the ordinary brothers were recruited for the study. Unlike the ordinary brothers, the twins had responded to newspaper advertisements specifically asking for volunteers, which might have attracted gay pairs.

I suggested that Bailey reanalyze his data. This time he should see if the pairs of gay twins had other brothers who were gay, then do the same for the pairs that had one gay twin and one straight co-twin. If Bailey's hypothesis about genetics was correct, the brothers of gay/gay twins should have been more likely to be gay than the brothers of gay/straight twins. Sure enough, when Bailey ran the numbers again, he found that the brothers of gay twins had a 22 percent chance of being gay compared with a 4 percent chance for the brothers of the gay/straight twins.

I still was concerned about the low, 9.2 percent, rate of homosexuality found in the brothers who weren't twins. Why wasn't it 22 percent as for the fraternal twins? I asked Bailey if anyone had done a study that ignored twins altogether and looked only at the ordinary brothers of gay men. In fact, Bailey's coauthor, Pillard, had performed exactly such a study in 1985, together with James Weinrich. Their study was extremely careful to avoid any volunteer bias; the advertisement for subjects did not even mention the word "sexuality." They found the brothers of the gay men had a 22 percent chance of being gay—the same rate found by Bailey in fraternal twins—whereas the brothers of straight men had only a 4 percent chance of being gay. These results gave me more confidence in Bailey's numbers.

Bailey later studied lesbians and their twins and sisters. The rates were almost identical to those for the men, which suggested a strong genetic link for women as well as men. Those results raised another question: Did male and female homosexuality run together in the same families, or were they distinct? Here the data were murky. Bailey found that the brothers of lesbians had a slightly elevated chance of being gay, but the increased likelihood was not significant. Conversely, Pillard and Weinrich had found that the sisters of gay men were more likely than average to be lesbian, but the elevation was less impressive than for brothers. These results seemed to indicate that male and female sexual orientation were at once distinct and partially overlapping, but there was not enough evidence to say if the cause was genetic, environmental, or some combination.

Bailey left me confident I could do a valid study to determine if there was some genetic basis for sexual orientation. The place to start, as with all studies in the field of genetics since the first attempts were made, more than 130 years ago, was with families. Because genes are handed down from parents to children, any trait that is linked to genes will run in families. So if being gay is at least partially genetic, it must run in families. To find out, we would need large families with more than one gay man, preferably gay brothers. We first looked only at men, to narrow the focus and because it was unclear if the same genetic mechanisms were at work in both men and women. The study of lesbians would come later. Once we found related men who were gay, we would study their chromosomes in the hope of finding a common area, or genetic marker.

AIDS, ALCOHOL, AND MENTAL HEALTH

Since I was going to collect such a large distinct group, in this case gay men, and I was going to study their sex lives in depth, analyze their family histories, and map their genetic makeup, it was important to search for other things at the same time. The obvious area was AIDS, which was found at higher rates in gay men than in heterosexual men. During my library research I also found studies suggesting that depression and alcoholism were more common in gay men and women, and a genetic study would offer a chance to test that hypothesis as well. The reason for bundling several projects together was to save time and money. We were going to collect a large amount of data on a controlled group of subjects, so it made sense to get as much out of them as possible.

I definitely wanted the study to have a medical segment to track the progression of HIV, the AIDS virus. In the United States, the population hit hardest by AIDS is gay men, who account for more than 50 percent of all cases. In some major urban centers, such as Los Angeles and New York, up to 20 percent of the gay male population is infected by the virus. One of the unanswered questions about the disease is why different people respond differently to the HIV virus. Some people who were infected at the beginning of the epidemic, more than fifteen years ago, still are perfectly healthy. Others became ill and died within a few months of their initial infection.

What is responsible for this striking variability in the progression of AIDS? There probably are many factors, including the particular strain of virus contracted, subsequent encounters with other microbes, and the general health of the infected person, but one possibility that largely had been overlooked is that some people are born with genes that restrict the ability of HIV to cause disease. If such genes could be isolated, genes that essentially protect the body from AIDS, it would offer a whole new way of approaching a cure for the disease.

Some of the best work on AIDS had been done right on the NIH campus, which was another advantage of working at a place with a large budget for research and the atmosphere of a university. Instead of plowing through dry research papers, I could walk across the grass and visit Michael Polis, an expert clinician at the National Institute of Allergy and Infectious Disease, and the epidemiologist William Blattner.

After talking with Polis, Blattner, and others, I was convinced the best way to study the progression of AIDS was to focus on one particular outcome of HIV infection, namely Kaposi's sarcoma, an unusual sort of cancer that usually starts on the skin but can spread rapidly, with lethal effects, to all parts of the body. Until recently the cancer was found only in elderly Ashkenazic Jews, Mediterranean men, and Central Africans. Then it began killing gay men with HIV.

I knew firsthand about the spread of Kaposi's sarcoma. One Wednesday night, a regular member of my bridge group, a 34-year-old man who worked at the Labor Department, showed up with what looked like a black-and-blue bruise across his nose. I didn't say anything, but I knew the discoloration on his face was a Kaposi's lesion. Normally an aggressive bidder and a sharp player, the man played sluggishly that night. Six months later, I attended his funeral.

Focusing on Kaposi's made sense for two reasons. First, Robert Gallo, who also works at NIH and is perhaps the world's best known (and most controversial) AIDS researcher, had speculated that the growth of the malignant cells in Kaposi's might be influenced by testosterone, the male sex hormone. This made my ears perk up, because several early researchers had speculated that testosterone and other sex hormones played a role in sexual orientation. If this were the case, the search for "gay genes" and "Kaposi's susceptibility genes" might lead in the same direction.

The second and more compelling motivation was a suspicion that the susceptibility of gay men to Kaposi's was influenced by their histocompatibility antigens. These are gene-produced proteins that are found on the surface of blood cells and that play a major role in the immune system. Previous studies of genetically unrelated individuals suggested that people with one type of histocompatibility antigen were more likely to get Kaposi's than were those who had another type.

One of the scientists involved in this research was Steve O'Brien, who works nearby at the Frederick National Cancer Research Center, which is funded by the National Cancer Institute, where I work. O'Brien told me that his group was doing a very large study of the role of genes in susceptibility to HIV. Their strategy was to get blood samples from different groups of people—gay men, Haitians, IV-drug users, and hemophiliacs—with varying HIV outcomes, including Kaposi's. They planned to screen hundreds of different genetic markers to see if any of them correlated with the viral outcome. This would be a gigantic effort costing many millions of dollars, but if they could find a gene that protected people from Kaposi's and other outcomes of HIV, it would be a revolutionary breakthrough.

But I detected a weakness in the experimental design. Since the study relied on genetically unrelated subjects, it would be difficult if not impossible to spot genetic links. How would they know which genes were linked to trivial differences in people and which ones were linked to disease?

This is why studies of genetics usually use families instead of individuals. By looking at the genes of the parents it is possible to predict the probable genetic makeup of the child, which makes it possible to isolate specific traits. When I raised this point with O'Brien, he said he would love to use families, but that it would be very difficult to find families with more than one person infected with HIV or at risk.

I told him I was planning to collect just such a group in a study of families with several gay men. I promised that if we found any families with two or more gay men who were HIV positive, we would share the data with his lab. In return, if he detected any promising genetic markers for Kaposi's or other diseases associated with AIDS, we would test our families for the markers.

Unlike O'Brien's study of unrelated individuals, I wanted ours to use gay brothers who were HIV positive and had the skin cancer. Our hypothesis was never that there was a gene that only gay men had that caused Kaposi's. Since other groups also got the cancer, we suspected some other factor was involved with gay men, such as a microbe transmitted during sex. Even if there were a gene that made people susceptible to Kaposi's, I knew it would be tough to pinpoint. To find a specific gene without having to check each of the 100,000 genes in the body, a shortcut was to look at specific regions of chromosomes. The only way to compare regions was to look at people related by blood.

The final areas of study that looked promising were mental health and alcoholism. In our culture, gays often are portrayed as depressed and dejected, alcoholic, and even suicidal. As one of the characters in the play *The Boys in the Band* put it, "Show me a happy homosexual and I'll show you a gay corpse." The research I found suggested that the association between homosexuality, alcoholism, and depression may not be purely fictional. Several studies had shown that gay men and women were two to five times more likely to have a problem with alcohol or drugs than their heterosexual peers. Others had claimed that gays were more likely than straight people to commit suicide and that gay teenagers accounted for a disproportionately high percentage of youth suicides.

There were serious problems with these studies, however. Most of them were conducted during the 1970s, when homosexuality was stigmatized even more than now. At that time, the easiest place to find gays was in gay bars, which probably weren't the best places to do unbiased studies of alcoholism. Finally, the studies avoided the question of whether sexual orientation and mental health had a physiological link, or if it was the societal disapproval of gays, and their feelings of guilt, that made them depressed.

So the mental health portion of our study was designed with two purposes. First, we wanted to determine if the connection between homosexuality, alcoholism, and depression was real or a myth. By comparing gay men with their heterosexual brothers, we hoped to study pairs of people with similar upbringings whose only apparent difference was sexual orientation. The second objective was to identify the genes that underlie these traits, especially alcoholism. Previous studies of twins and families had suggested that such genes probably existed, but no single gene had been identified.

This part of the study would involve pairs of alcoholic gay brothers. Here again, as with the Kaposi's study, the emphasis was on the men being brothers, not on them being gay. Our hypothesis was not that the "alcoholism gene" would be especially prevalent in gay men but that it could be detected by studying any group of alcoholic brothers. We wanted to look at families that had numerous alcoholics in different generations. If the theories about gays and alcohol were true, then families of gay men should have a relatively large number of alcoholics.

Around the time we were preparing our study, a research group in Texas claimed to have found a specific gene, coding for the dopamine D2 receptor, associated with alcoholism. I didn't know what to make of the study, but I was encouraged because it was another potential connection between genes and behavior. The mention of this specific gene was intriguing because dopamine is a chemical in the brain thought to play a key role in rewarding dependence and pleasure. I could easily imagine how changes in these pathways might make a person dependent on alcohol.

I bounced the finding off two of the NIH's experts on psychiatric genetics. Unfortunately, I got opposite answers. David Goldman, who's in the National Institute of Alcoholism and Alcohol Abuse, told me his group had tried to replicate the Texas finding and failed. Goldman, who is high-strung and excitable even on quiet days, worked himself into a lather pointing out the weaknesses and inconsistencies of the Texas work. But another colleague, Elliot Gershon of the National Institute of Mental Health, was impressed with the Texas study. His group planned to search the entire D2 receptor gene for mutations in its coding sequences. He said it was too early to tell if the genetic link to alcoholism was real, but he thought it was worth trying to replicate the Texas work.

To my mind, the reactions of Goldman and Gershon to the "alcoholism gene" suffered from the same problem as Steve O'Brien's study of Kaposi's sarcoma: too much reliance on individuals instead of families. Maybe I was spoiled by working with yeast cells that multiplied faster than anyone could count, but I thought I could collect families with lots of gay men. Perhaps the same families would have lots of alcoholics.

As I had done with O'Brien, I struck a deal with Goldman and Gershon: If they discovered interesting mutations in the dopamine receptor gene or any other gene that might be involved in alcoholism, I would test for it in my families. In return, I would make my samples available to them. I was becoming increasingly popular around the NIH among people who wanted to dip into the sample I hoped to collect, because gay men are a rich vein of medical information for scientists studying everything from AIDS to Kaposi's sarcoma to depression and alcoholism to sexual behavior.

THE HUMAN GENOME PROJECT

All the topics I was researching, from sexuality to AIDS progression to mental health, are what scientists call "complex." That is, none is caused by any single factor, genetic or otherwise. Even if heredity does play a role, there is no way to predict—without looking at the DNA—how many different genes are involved or what they do. What made it possible even to think about identifying single genes for such complex characteristics was the rapid advance in genetic technology.

When I did my training as a Ph.D. student at Harvard during 1972–77, and later as a postdoctoral fellow in Philip Leder's laboratory at the National Institute of Child Health and Human Development, mapping human genes was very difficult and tedious. In a very small number of cases, two genes that coded for obvious physical traits, such as hemophilia and color blindness, happened to be close together on a chromosome and their positions could be deduced from family studies. But in most cases, the only genes that could be found were the ones that already had been isolated or cloned. Since it was necessary to know what a gene produced to be able to find it, searching for genes that might not even exist was like flying blindly around the solar system, hoping to bump into an alien spaceship.

Starting in 1990, the scientific community and the U.S. government initiated a fifteen-year, $3 billion plan called the Human Genome Project. The goal is to precisely map all three billion base pairs, or bits of information, that make up the human genome, which is the complete complement of genetic information in a single person. The first step was developing genetic markers, which are signposts that indicate the exact position of a piece of DNA on a chromosome. Scientists also were developing ingenious ways to use these markers to map genes even when it wasn't known what the gene did or where it was located on the genetic map. They already had used these methods to map genes for several important diseases, such as cystic fibrosis, Huntington's chorea, and one type of colon cancer.

At the same time, theoreticians were thinking about how these markers could be used to map genes that contribute to complex traits—traits in which many different genes are involved or in which genes are only part of the story. The technology involved in the Human Genome Project is continuously improving, so eventually we should be able to isolate the gene for anything that is genetic or even for something that is influenced by genes some of the time. Someday we will be able to determine the exact DNA sequence not just of one person but of entire groups of people who are similar, or different from one another, in any given way.

I knew these advances in technology could help us while we were searching for a "gay gene," or for other genes involved in Kaposi's, alcoholism, and mental health. With luck, I thought, new breakthroughs would give us still more tools to use. I was especially impressed by Mary-Claire King's work on breast cancer, which in a small percentage of families is caused by an inherited gene defect. Usually breast cancer is "sporadic," which means the gene must be spontaneously altered or combined with diet or estrogens to cause the cancer. Despite the complexities involved, King and her colleagues had convincingly mapped a contributory gene on chromosome 17, a discovery that was confirmed by several other laboratories. Many scientists speculated that a similar approach could be used for other common, partially genetic traits. Perhaps homosexuality was one of these.

I finally had all the information and guidance I needed. What I had learned during my six-month crash course was that there was good reason to suspect genes somehow were involved in sexual orientation. Also, there was a reasonable likelihood that other genes were involved in HIV progression, Kaposi's sarcoma, and alcoholism, all of which were believed to be found at increased levels in gay men. From those two hunches grew the kernel of my idea: a full-scale exploration of the genomes of gay men. I knew that identifying individual genes for complex traits would be tough, but I also knew that if nobody looked, they'd never be found.

I had gone as far as I could on my own, reviewing the published literature and talking with my colleagues about their data. The problem wasn't analyzing the data; the problem was that there wasn't enough of it. No amount of time in the library or talking to more experts was going to produce any, either. The only way to get data was to run experiments.

Chapter Two

THE STUDY AND THE TEAM

Despite the freedom I enjoy as a senior staff member at the National Institutes of Health, I couldn't just run off and work on whatever I found interesting or challenging. First there had to be a proposal, known as a protocol, and peer review. Then funding had to be approved and a research team formed. Finally, I needed volunteers to test.

When I mentioned I was thinking about working on sexual orientation and sexually related health matters, friends warned me that I was walking into a political minefield. To me the science seemed reasonable and the medical relevance obvious, but I realized from the start that people are not always rational when it comes to sex. That was why I wanted to be sure our proposal was sound not only scientifically but also medically, economically, and even politically. It also was the reason I wanted to organize a team of scientists with solid reputations and proven expertise to help me.

PROTOCOL #92-C-0078

I spent several weeks at my computer, preparing the protocol, outlining what we planned to do and why. I estimated the study would last three years and would involve 230 subjects. By the fall of 1991 I had drafted the basic plan for the study, which was designated NIH Clinical Protocol #92-C-0078. The title was long but descriptive: "Genetic Factors and Interrelationships for Sexual Orientation, HIV Progression and Kaposi's Sarcoma, Alcoholism and Related Psychopathology, and Histocompatibility Antigens."

Taking the advice of colleagues who had run their own protocols involving human subjects, I assembled an advisory committee to help with the legal and ethical issues and any problems that might arise. I included two bioethicists from the NIH, a lawyer from the NIH director's office, the president of the Society for the Scientific Study of Sex, a minister from a local Presbyterian church, the medical director of the Whitman-Walker gay and lesbian health clinic, the director of the Federation of Parents and Friends of Lesbians and Gays, and an AIDS and sex researcher, Jim Weinrich.

Colleagues also helped teach me about the nitty-gritty of recruiting volunteers, where to interview them, how to protect the rights and confidentiality of the volunteers, and how to cope with medical emergencies. At that point, I didn't even know how to draw blood, so I would need to take a one-week course.

As with all NIH studies involving more than one institute, the first hurdle was an evaluation by the clinical directors of the four participating centers: the institutes for Cancer; Child Health and Human Development; Allergy and Infectious Diseases; and Alcoholism and Alcohol Abuse. After suggesting a few minor revisions, all four clinical directors signed off on the protocol. Next came the Cancer Institute clinical review subcommittee, which includes physicians, nurses, bioethicists, and people from the surrounding community of Bethesda who examine every protocol involving research on human beings. Their role is to guarantee the scientific merit of the project and to protect the rights and health of the people being studied.

I was scheduled to present my case directly to this group on 18

November 1991. I was nervous about the interview because I didn't have any idea how the members of the committee felt about the general topic of homosexuality, a subject about which almost everyone has an opinion. When I looked over the list of members, I was struck that they were such a broad and diverse group. Some had devoted their entire lives to fighting cancer; others had been involved with churches and matters of the spirit. Some were well-known researchers, while others had backgrounds with absolutely no relation to science. I decided it was best to be straightforward and honest with them and not try to fudge any of the details. The work was going to become known eventually, so it didn't make sense to try to hide anything. I did dress for the occasion, exchanging my usual lab wear of jeans and a T-shirt for the unofficial NIH business uniform of khaki pants, a button-down oxford-cloth shirt, and a tie.

I arrived a few minutes before my scheduled appearance and was told to sit in a waiting room. In the next room members of the subcommittee were sitting at a large table, reviewing protocols and interviewing candidates about their research proposals. Two physicians who were next in line to go before the committee already were in the waiting room when I got there. They had a fat, 200-page proposal and an armload of supporting documents. They told me they wanted to give a novel anticancer drug to thousands of patients and track the results over a long period. I sheepishly fingered my own skinny protocol and wondered how the committee would react. All I wanted to do was interview people about their sex lives and take a few tablespoons of blood. I wasn't sure if the committee would see the plan as trivial and stupid, or perhaps as so outrageously unconventional that they would jump out of their chairs in alarm. On the plus side, I thought to myself, the medical benefits were obvious, if we could find a genetic link to HIV progression, Kaposi's, or alcoholism. Besides, the subjects of the study would run almost no risk of harm.

Finally they called my name and summoned me into the conference room. I stood before the group and read a brief summary of my plan, which they had reviewed before the meeting began. When I finished, the committee members had only two concerns. The greatest was that I would protect the confidentiality of the people involved. I explained that the volunteers would be told that the results of the study would be published, but all names would be kept secret. Rather than being worried about me revealing the results to the public, the committee was concerned I would reveal private information from one family member to another. They worried, for example, that I would tell unknowing parents that their son was gay. I explained that I would contact family members only with the express, written permission of the volunteers, and only after giving them time to talk with their parents or siblings. That would prevent anyone from being surprised or offended. A second protection would be a standard consent form that everyone involved in a study like this must sign.

The other question, from one of the medical doctors, was more revealing: "Hasn't it already been determined that there are genes that influence homosexuality?"

"Not that I am aware of," I said. He seemed to think the topic was so obvious that it must have been done before. I had been just as surprised myself when I started looking into the subject and found how few real experiments had been done.

A few weeks later I received a fax from the committee. If I agreed to add an additional consent letter for contacting family members, my protocol would be approved. That seemed like a good idea, and I drew up a sample consent form and sent it to the committee. A few days later they signed off on the protocol.

Once the project was headed toward approval, I needed funding. My laboratory is automatically funded annually for routine operations, and the only big item we needed to buy for this study was a new microscope, which could be used for many other things if the sex research didn't go as planned. We also wanted to bank our blood-cell samples to store the genetic information in a permanent repository. If we discovered some interesting gene down the road, we wanted to be able to go back and look at the original samples. To store the samples myself at NIH would have required a full-time technical assistant, and because getting a new person hired is a bureaucratic challenge, I was afraid my request for an assistant would be rejected. Even if I got a new person, we didn't have a lab in my area that was equipped to handle HIV-infected blood, so I searched until I found a commercial lab in New Mexico that was able to store the samples cheaper and more reliably than we could. Because the bank was a contract facility not controlled by NIH, there would be no question that other groups would have access to our biological samples to test our results or to expand on our work.

The funding proposal was modest and caused barely a ripple. The NIH's 1992 budget of $11 billion included my lab and staff plus Protocol #92-C-0078 at a cost of $75,400.

With the science approved and the funding assured, the only remaining hurdle was political. I knew this wasn't one of the usual studies of human genetics searching for a rare trait such as polydactyly or ataxia telangiectasia or some other scientific mouthful that was too technical to cause a stir. This was a topic everyone could understand, and while I was working on the proposal, it seemed the entire nation was up in arms about gays in the military, and congressional hearings were being held on Capitol Hill, just a few miles from the pleasant, grassy campus of NIH. There were few aspects of homosexuality that were being discussed with anything like a cool, detached scientific view, and from the tone of the debate, I knew the right wing would question our work because they think being gay is a "lifestyle" that people choose, not something genetic. The left wouldn't be happy, either, because they would worry about homosexuality being classified as a genetic defect that could be "cured," instead of as a normal human variation.

When I first discussed the proposal with my boss, a French scientist named Dr. Claude Klee, she put it this way: "In France we have no problem with such business, but in America you are such prudes, no?"

Then she looked at me over the top of her glasses and told me in a heavy Marseilles accent, "Whatever you do, stay away from the psychiatrists."

As the chief of the Laboratory of Biochemistry, Klee was good at both science and administration. She was seasoned enough to know that the wisest thing to do with a proposal like mine was to pass it along to her supervisor, the director of the Division of Cancer Biology and Diagnosis at the Cancer Institute. The director said he liked the cancer connection in the study of Kaposi's sarcoma, but he decided to kick the whole thing up yet another level, all the way to the director of the National Cancer Institute. Since the law of bureaucracy is stronger even than the law of gravity, the proposal continued upward until it could get no higher: the desk of the director of the entire National Institutes of Health, Dr. Bernadine Healy.

I knew that under Healy, a conservative political appointee of the Bush White House, funding for a previously approved national sex survey was cut, so I would not have been surprised if our proposal had been stopped. I saw a copy of a memo she prepared for her boss, Assistant Secretary of Health and Human Services George Mason, entitled "Upcoming Research with Possible Sensitive Issues." The memo did not ask for permission to do the sexual orientation research but simply stated it would start soon and that Mason should be "aware of it should questions arise."

The only question from on high that ever made it back down to my level was from Mason. I was told that the assistant secretary wanted to know why we had lumped three different topics together. That was a reasonable question, and I prepared a response that explained how we could save money by using a single group of volunteers, in this case gay men and their families, to study sexual orientation, HIV and Kaposi's, and alcoholism. Apparently that was enough to convince Mason, because the study went forward.

I wanted a copy of all these high-level approvals for the record, but I'm not sure they ever were put on paper. Even high-level approval was not a 100 percent vaccination against political problems, however, and there were other approved projects that had been killed after unfavorable attention in the Congress or the media. Fortunately, our troubles didn't start until after we had completed the research. At that point, our opponents made up in ferocity what they had lacked in timing.

We still needed a place to do our research. The bench work would be done in my laboratory, but I needed a separate and more private space in which to interview the volunteers. NIH might be big, but every inch is spoken for—especially in the Clinical Center, which is the logical place to see volunteers. One of the major institutes offered space in its clinic, but only if one of its staff members, a psychiatrist, served as the medical doctor assigned to the project. I must have temporarily forgotten Dr. Klee's warning about psychiatrists, because I didn't object to that arrangement. Everything was fine until the psychiatrist informed me that she would not do physical exams on gay men because she was afraid of "catching AIDS." A doctor who wouldn't touch our subjects was not going to be much help, but I wasn't sure how to get out of the commitment.

Eventually we found space in the Inter-Institute Genetics Clinic and convinced Lawrence Charnas, a bright young neurologist, to be the medically responsible investigator for the project.

THE TEAM

Now we had approval, funding, and a place to work. But there was no way I could do a project like this myself; I needed a research team. I planned to dedicate the majority of my time during the next several years to this study, mostly recruiting and interviewing several hundred subjects and taking samples of their DNA. Other people would be needed to help process the samples and analyze the results.

In addition to the time I would be spending on the sexual orientation study, I had several other projects in the works. I confess this was to protect myself. Far from being sure we were going to find a genetic link to homosexuality, I worried that the whole enterprise would collapse for lack of evidence. I had seen too many other experiments flop to be overconfident, so I also planned to work on two projects that were more conventional. One involved genetic studies of courtship behavior in fruit flies, and the other involved molecular studies of the sexual differentiation of the hypothalamus region of the brain in rats. The study involving people probably would take years to reach a conclusion, but in the meantime I could get some quick results from the fruit flies and rats. Besides the three research projects, I had the normal administrative and academic responsibilities of running a lab at the NIH. All this meant I wouldn't have much time for the detailed molecular chores that were at the core of any human genetics project.

In early 1991, when I first started thinking seriously about the project, there were ten scientists with me in my laboratory. The one person I really needed was Stella Hu, who had been my technician for four years. She was born in 1932 in Nanking, China, and her well-educated parents had sent her to a missionary school. She earned her B.S. and M.S. in biochemistry in Toronto and had long experience as a technician. She is the most tenacious bench worker I've ever seen in action, and if a technique doesn't produce results the first time, she'll keep changing it until the experiment works perfectly.

Stella was perfect for the homosexuality project. There was only one problem: She was absolutely, completely opposed to the idea. More than that, she was horrified. I quickly learned that she barely could say the word "homosexuality." When she tried, all 5 feet 2 inches sputtered: "I couldn't possibly work on *that!* I'd be too embarrassed to give a seminar."

I was Stella's boss and could have ordered her to do the work or leave the lab. I didn't want to lose her, though, and the truth was I needed her help. We compromised by letting her focus on the part of the study dealing with HIV progression, Kaposi's, and alcoholism. Someone else would search for the "gay gene." Of course all of our families would have multiple gays, but this was acceptable to Stella as long as there was some obvious connection to medicine. I didn't argue, because I was relieved to have her stay. I did casually lend her a few scientific articles about homosexuality, just to pique her interest. After that, I didn't say another word. Better not to upset her any more than necessary.

I also hired a woman named Nan Hu (no relation to Stella) because of her many years of experience analyzing chromosome structure in cancer patients. Nan Hu was born in China, like Stella, but didn't have any strong ideas one way or another about homosexuality. To her it was a "characteristic" rather than a "disease." The third scientist to join the team was American-born Angela Pattatucci, who had had a varied career that included high school teaching and nursing and had recently completed a Ph.D. in molecular genetics at the University of Illinois.

Three other scientists would play important roles in the lab, although they were not involved in the human research. The first, Roger Gorski, is a professor and former chairman of the Department of Anatomy at UCLA who was at our lab on sabbatical to learn molecular genetics. Eighteen years before, Gorski had discovered one of the most significant sexual differences in the rat brain, and recently he and his colleague Laura Allen had found a subtle difference in brain structure between heterosexual and homosexual men. Jin Zeng and Bennet Prickrill, both postdoctoral fellows, rounded out the team.

REAL PEOPLE

Our theories didn't mean anything until we tested them on real people. In the past, I had never had trouble recruiting excellent subjects for my experiments. I just walked over to the freezer, took out some yeast cells, and put them in broth overnight. The next morning I had more "volunteers" than I could use.

Finding the right gay men was going to be harder. In principle, I might have collaborated with an AIDS center that had blood specimens and medical records for large numbers of homosexual men, but I needed more information than typically is found on a medical record. I wanted to know about my subjects' sexual thoughts, fantasies, behaviors, and attitudes and how these had unfolded over the course of their lives. Even more importantly, I would need to meet and obtain blood specimens from their relatives—their siblings and parents and, in some cases, their aunts, uncles, and cousins. These types of samples were not available in any freezer. The first thing I did was to go out and buy some new clothes. I would be spending the first few months out of the lab and on the road.

Random Subjects

We actually needed two quite different groups of gay men. The first would be a random group used to determine whether being gay runs in families. In this sense, "random" doesn't mean a completely random-probability sample of gay men, but a sample randomly selected without bias with respect to any gay relatives. Accordingly, it was important that we not know in advance about the families of the subjects or whether they had gay relatives. Likewise, they should not know about the purpose of our study. We were careful never to mention the words "genetics" or "biology" to them, because we didn't want to deliberately attract men with lots of gay relatives.

We started the search at the HIV Clinic of the National Institute of Allergy and Infectious Disease. Because an important aim of the project was to identify genes that modify the progression of AIDS, we needed access to HIV-positive gay men with detailed medical records and the possibility of follow-up studies. Although outpatients went to NIH primarily for treatment of their HIV disease, they were routinely offered the opportunity to participate in research programs covering a range of subjects. For us, the broad range of research at the clinic meant that if we advertised a protocol on "Human Sexual Development," it would not seem terribly out of the ordinary, and it might attract a reasonable number of participants. Another consideration was that being HIV positive, these men already had given their doctors very detailed histories of their sex lives.

To find additional subjects for this random group I went off campus to the Whitman-Walker Clinic, which serves gay men and lesbians. Located within walking distance of Dupont Circle, the hub of the capital's gay community, Whitman-Walker provides a range of health services, including counseling and treatment programs for alcoholics and substance abusers. Another source of participants was the Triangle Club, a private center that offers gays and lesbians the 12-step programs of Alcoholics Anonymous and Narcotics Anonymous. Because the hunt for genes that influence addictive behaviors was an important component of the study, the clinic and the Triangle Club were natural places to look for subjects.

The final source of gay men for our random group was Emergence, an organization formed by gay and lesbian Christian Scientists. The attraction of this group was its small size, which allowed us to interview every member who attended two or more meetings over the course of four months. That meant we would not be picking only the "activist" members or the eager volunteers; we got everyone and thus avoided any bias in the selection process.

Brothers and Families

Besides this random group of gay men, we needed a second group of people, consisting of families with more than one gay member, for our DNA linkage studies. The obvious place to start was with Parents and Friends of Lesbians and Gays (PFLAG), a support and advocacy group. The original notice, which was mailed to nearly 20,000 of the group's members in the spring of 1992, called for families with "three or more gay men or lesbians." This is the classic way of doing genetic studies. The favorite subjects are extended families containing a large number of the target population, in this case gays. The search for the gene that causes Huntington's chorea, for example, was based on a huge family in Venezuela that had suffered numerous cases of the disease for generations.

The method didn't work well for us, however. For example, we found one family with three gay brothers and two lesbian sisters. From what I had read, homosexuality in males and females usually runs in separate families, so one family is not likely to have both gay men and lesbians. If this weren't a typical family, we might not find a typical gene. Another family had two generations of incest, including a father and son, and other severe emotional problems that made it unusual and not representative.

Although the mass mailing produced 15 usable families, we decided by the spring of 1992 that focusing the DNA linkage study on a few large families was a mistake. Instead, we would look at pairs of gay brothers. This is a technique that often is used in genetic research because siblings share 50 percent of their genetic material. We started the search again, placing an ad in the gay newspapers in Baltimore and Washington, D.C., that read:

RESEARCH VOLUNTEERS NEEDED

Gay Men—Do You Have a Gay Brother?

We are seeking gay brothers for a basic research study sponsored by the National Institutes of Health. If you and your brother agree to participate, you will be asked to complete a

> short interview and donate a blood sample at the NIH in Bethesda, MD. Transportation costs will be provided. Your identity and participation in the project will be kept strictly confidential, and people with HIV or alcoholism are welcome.
>
> For more information contact Dr. Dean Hamer, . . .

The only incentive we offered was free airfare to bring the brothers to Washington. The result was that most of the people who responded were local gay men whose brothers lived out of town. The out-of-towners could stay with their brothers, a convenient arrangement because the NIH provided volunteers only $45 a day for living expenses. Unfortunately for me, the ones willing to come to Washington naturally wanted to come during holidays. Others didn't want to come at all, so I traveled across the country to them. The advantage of that was I could see the families in their natural habitat. In the end, 23 pairs of brothers volunteered in response to the ads.

A Select Group

Eventually we had 114 gay men for primary volunteers, or index subjects. They included 76 "random" subjects whose family histories were not known and who would be used to see if homosexuality ran in families and appeared to be inherited. Another 38 gay men were chosen precisely because each one had a gay brother. Their DNA would be studied to see if we could find any common areas that indicated a genetic link to their homosexuality. We also interviewed 142 relatives of both groups, 99 male relatives and 43 female relatives. The relatives would be used to draw the family trees of sexual orientation and to give us reference subjects with whom to compare the gay men.

When I scanned the list of subjects we had collected, it definitely did not "look like America." It looked more like the membership list of an exclusive downtown health club or the Young Lawyers of America. Our subjects were not even representative of the larger gay community. Then again, they weren't meant to be. They were overwhelmingly white: 92 percent Caucasian (not Hispanic), 4 percent African-American, 3 percent Hispanic, and 1 percent Asian. They tended to be highly educated, well-to-do, and nearly middle aged. The average educational level was 15.5 years, the average income was greater than $40,000 per year, and the average age was 36 years. Most importantly, they all were open about their sexuality; otherwise they would not have volunteered for the study. Most but not all were liberal Democrats who supported Bill Clinton for president; that was irrelevant to the larger study, but it did allow us to do a clever control experiment that will be described later. This demographic group of middle-class whites is not unique to studies of gays. Most scientists would agree that, aside from their sexual orientation, these are the type of people who typically volunteer for any kind of scientific study that doesn't pay participants.

The group would have been hopelessly biased if we had been trying to answer a question such as, What is the level of disposable income among gay men? But for our purpose—to find out whether genes influence sexual orientation—they were perfect. That is because income, education, political affiliation, and the like are not, at least as far as we know, inherited characteristics. The advantage of working with such a sophisticated group of subjects was their openness about sexuality, a characteristic that would help rather than hinder our research.

We now had a large number of willing volunteers. The next step was to systematically classify them according to sexual orientation and other factors. This sounds easy enough, but it proved a challenge to take something as difficult to quantify as sexuality and make it fit into the neat charts and graphs so loved by scientists. This part of the study was unlike anything my lab had ever tried before, and instead of staring at fruit flies or yeast cells, we were going to explore the most private thoughts and actions of real people.

Chapter Three

WHO'S GAY?

Just what does it mean to be gay? Is it a matter of attraction, sexual behavior, or both? Is it defined by who people have sex with, or who they would like to have sex with? These are questions that have been debated by researchers, not to mention historians, sociologists, the courts, and most recently the Pentagon. We decided to operate on the general principle that the more we knew, the better. Our aim was to learn as much as possible about all aspects of a person's sex life, both psychological and behavioral, past and present. We began calling our volunteers to arrange the lengthy interviews that would start the process of finding out who was gay, who was straight, and who was in between. There were problems, as expected, and a few concerns about the science of sex but also some very interesting results.

Since many of the subjects had responded to requests for gay men, we had little trouble determining their sexual orientation. We needed to document this, however, with detailed questions about their sex lives. With the relatives of the gay men, the challenge was greater. In some cases, for example, a gay man identified a cousin as gay, but the cousin didn't consider himself gay. Or there was the cousin who had slept with a man once but then married and had children.

Most people would say without hesitation that they know whether they themselves are homosexual or heterosexual. Most people also would say they have a good idea who is gay and who is straight in their own families. The same is true for close friends. Even walking down the street, many people would claim to be able to point out at least some of the "obviously" gay—or straight—men and women passing by. So measuring sexual orientation is something we do informally all the time. For the study, however, we needed to do it in a way that was scientifically sound, quantitative, and could be replicated by other investigators.

THE MEANING OF PHENOTYPES

The challenge of scientifically measuring a trait is not unique to studies of sexuality. In fact it gets to the very core of genetic research, which revolves around the connection between genotype and phenotype. The genotype is simply the precise makeup of a person's inherited factors, the genes. Phenotype describes a person's observable characteristics, whether anatomical or psychological. Genotypes can be precisely defined because they consist of DNA molecules whose information content is specific and unambiguous. They can be mapped, diagrammed, and classified scientifically. Phenotypes are harder to pin down. For example, everyone could agree on the differences between the phenotypes of brown eyes and blue eyes. But ask who is "tall" and who is "short," or who has "light" skin and who has "dark" skin, and opinions will vary.

When we want to classify a trait such as sexuality, the definition of phenotypes becomes far trickier than with eye color. In part this is because sexuality encompasses so many different aspects of a person's physical, mental, and emotional makeup. A second complication is that these characteristics are not rigidly fixed during the life span. Most people do not have the same sexuality at age 50 as they had at age 20, and sometimes a person's desires can change within a period of a few days or even minutes. Because of the fluid nature of sexuality, it is important not only to isolate specific traits but also to know when they first appeared.

Yet a third complication is that sexuality carries a heavy load of nonscientific baggage, such as morality. Think, for example, of the phenotype of "promiscuity"—it probably means something different to a Baptist Sunday school teacher than to her teenage students. This complexity makes it important for sex researchers to clearly define the phenotypes they are studying.

THE INTERVIEW

Our assessment of sexuality was based on a structured interview that was conducted in private and usually lasted from one to two hours.* The areas covered were those typically used in modern sexuality research.

One advantage to living in a society bombarded with TV talk-show programs that revel in the most secret, private sexual activities is that people have become accustomed to talking about sex. Putting a microphone in front of them only makes it seem more normal. Before I started the interview, and again immediately before the questions concerning specific sexual acts, I told the subject: "All the questions are optional. If you don't want to answer a question, just shake your head 'no' and I'll move on." In the end this proved unnecessary; of the 213 men who were interviewed, only one declined to answer all the questions. This was not surprising, given that all the subjects had volunteered to be in a sexuality study. Actually many of the HIV-positive men were surprised the interview was so "short." They had been asked far more detailed questions about their sexual behavior by their physicians.

The questions were the same for each participant, and the structure of the sexuality portion of the interview was as follows:

* The interview questionnaire is found in Appendix B.

I. CHILDHOOD
   A. First emotional and sexual attractions
   B. Sex play
   C. Abuse

II. ADOLESCENCE AND EARLY ADULTHOOD
   A. Puberty
   B. Dating, romance, and socialization
   C. Sexual activity
   D. Developmental profile, early adulthood to present

III. CURRENT (PAST YEAR)
   A. Frequency of sexual activities
   B. Types of sexual activities
   C. Emotional attachments

IV. PARTNERS
   A. Lifetime number (men and women)
   B. Past year number (men and women)
   C. Relationships (nature and duration)

V. QUANTITATIVE (KINSEY) SCALES
   A. Self-identification
   B. Attraction
   C. Fantasy
   D. Behavior

VI. SEXUAL ACTS

THE STORIES OF BRIAN AND MICHAEL

Most of the histories definitively determined whether the subject was gay or straight. For example, here are the stories told by two brothers from one of the families we studied extensively, whom we will call the Wilson family.

BRIAN WILSON: "Of course I can remember my first crush. It was a little girl down the street who had the blondest hair I've ever seen. I was really in love with that girl, but she wouldn't have anything to do with me. That was in second or third grade.

"My real interest in girls didn't begin until seventh grade, when we started having basement parties and we'd all go down to the basement, turn out the lights, and French kiss for three hours. If you didn't have a girlfriend you were out of luck. That was about the closest to an orgy I've ever been; I'm sure [my brother] Mike has a lot more experience on that front than I do.

"All through high school I had steady girlfriends. I guess I was considered a catch since I was on all the sports teams, but try as I would, none of them would go all the way. The only girl I did actually sleep with was definitely the town slut. Everyone knew she'd put out.

"Then I went into the military, and the very first night we had leave we all went into town looking for hookers. That was okay for a while, but it was awfully expensive for a private, and besides I would always feel guilty afterwards. I was glad to get out of the service and go back home.

"I met my wife, and it was love at first sight—which is kind of weird seeing as she's a small redhead and I usually like big blondes. We got married very quickly, and at first it was just great to have our own little house and come home and make love every night. But then she got a job, I got more and more busy with work, and eventually our son came along, and well, you know how it goes. Our love life really went down the tubes.

"These days we're getting along pretty good. We make love maybe once or twice a month. I think it would be more if it weren't for the kid and all. But sex is only a small part of my life. Family and work are just as important."

This interview led us to rate Brian as exclusively heterosexual. Next, we spoke to his brother Michael, using the same format and questions in the interview.

MICHAEL WILSON: "When I was about four or five, we had the house painted by a handyman. I guess it was summer, because he was working with his shirt off. I must have been turned on, because that evening I announced that I was going to marry him. My parents thought that was hysterical and told it as a big joke to all their friends.

"My first sexual contact was in the second grade, with a boy in my Cub Scout troop. We fooled around together a couple of times, but I was very afraid of getting caught—not so much that we were two guys, but because I was a good Catholic boy and wasn't supposed to do anything like that. I don't remember exactly what we did, but I don't think it was anything very serious.

"I reached puberty at the beginning of junior high, probably around 12 or 13. I remember my first ejaculation very well because it was a wet dream about a guy. I felt incredibly guilty about that. When I masturbated, I would always think about men, but I'd try to change my fantasy to a woman at the very last minute. Somehow I believed that if I was thinking about a girl when I came, it wasn't such a sin. I did date a few times in high school, but mostly it was group dates, going to the movies together, so then there was really no pressure on me to make out with the girls.

"Going away to college was quite a shock. I guess because it was so much more liberal than where I grew up. In a way it was kind of scary because there were so many women who were willing to have sex. I wasn't used to that. Anyway, I got a girlfriend, and we hung out together for about a year. We slept together, and physically it was okay, but I knew it wasn't what I really wanted. It was right around then that I discovered the gay bookstore scene and started having sex with men. I was still very afraid of people knowing, so I never went home with anyone. Too risky.

"I didn't really come out, even to other gay people, until 1975, when I got my first serious job. They didn't have a bookstore in that city but they did have a gay bar, so I started hanging out there on weekends. That's where I met Bill; we've been together ever since. At the beginning I'd still trick out every now and then, especially if Bill was out of town, but once the epidemic started, it just wasn't worth it anymore. I can't even remember the last time I slept with someone else.

"Now we have sex together about once or twice a week; more if we're on vacation, less if work is really bad or we're having a spat. I'm very content with my sex life with Bill. Maybe it's not as exciting or thrilling as it used to be, but it's sure better than the dating scene or being alone."

•

Based on the interview with Michael, we classified him as predominantly if not exclusively homosexual. For both brothers the ratings were obvious and likely would not be contested by other researchers.

THE KINSEY SCALE

In the case of the Wilson brothers, it was clear that Brian was straight and Michael was gay. We needed to determine if their clear-cut sexual orientations were the exception or the rule. Was sexual orientation, like eye color or being left-handed or right-handed, a discrete phenotype that comes in just a few, distinguishable classes? Or was it more of a continuous phenotype, such as height or intelligence, that ranges from one extreme to the other, with no obvious divisions between them?

To answer this question, it was important to use a quantifiable index of sexual orientation. We used the classical measure called the Kinsey Scale, which was developed by researcher Alfred Kinsey for his pioneering sex studies in the 1940s. Some of Kinsey's work reflects the time in which he lived and the infancy of sexual research, and a great deal of scientific research has been done since then that has improved on his work, but the scale Kinsey developed is still regarded as valid. The scale has seven levels, ranging from exclusive heterosexuality to exclusive homosexuality. We applied the scale to four distinct aspects of sexuality: self-identification, attraction, fantasy, and behavior. The way these scales work can be illustrated by the following interviews with the two Wilson brothers.

INTERVIEWER: "Now I'd like to ask you some questions with numerical answers from zero to six on what is called the Kinsey Scale.

"Zero stands for someone who identifies himself as exclusively heterosexual. One means a man is predominantly heterosexual but every once in a while is interested in other men. Two is a man who identifies as heterosexual but is attracted to or active with men more than just occasionally. Three is fully bisexual, meaning equally interested in men and women. Four is someone who is gay but is attracted to or active with women more than just occasionally. Five is predominantly interested in or active with men. Six is exclusively gay.

"How would you identify yourself overall on this scale?"

BRIAN WILSON: "I'm a zero, definitely heterosexual."

MICHAEL WILSON: "I guess I'd say five. Definitely gay, but not absolutely 100 percent."

INTERVIEWER: "Now I'd like to ask about sexual attraction. Here's another scale on which zero stands for exclusively attracted to females and six stands for exclusively attracted to males. Suppose you walk into a party, you notice various people, and you think to yourself, 'That one might be interesting to go to bed with.' What sex would that person be?"

BRIAN: "I recognize that some men are attractive that way, but only because a woman I've got my eye on goes for them instead of me. In terms of who I'm attracted to, it's strictly women. Zero again."

MICHAEL: "Well, if you're talking about sexual attraction, I guess I'd have to say six. I often think women are nice looking, but only in an aesthetic sense."

INTERVIEWER: "The next question is about sexual fantasy. When you imagine yourself having sex with someone, for example, when you masturbate, what is the gender of that person?"

BRIAN: "Why bother jerking off if you're not going to think about women? Give me a zero."

MICHAEL: "Definitely men. A six."

INTERVIEWER: "The final question is about behavior rather than thoughts. What mixture of men and women have you actually had sexual relations with?"

BRIAN: "I did have a guy come on to me in the army, but I've never actually done anything, so I guess it's zero again. Sorry to be so boring."

MICHAEL: "I had intercourse with my girlfriend in college, even though I'd go cruising the gay bookstore afterwards. So I'm not really a six. Is five and a half an okay answer?"

INTERVIEWER: "Yes, it is."

From these interviews, we assigned overall Kinsey ratings of 0.0 to Brian and 5.6 to Michael, fully consistent with the sexual histories they told us earlier.

CHECKING THE SCALE

The Kinsey Scale helped us classify our subjects as heterosexual, homosexual, or bisexual. But given the complexity and variability of human sexuality, one might well ask whether this pigeonholing of people into just three discrete categories is valid. In order for such a scheme to be scientifically meaningful, it must display three characteristics: consistency, reliability, and stability.

Consistency

The first requirement for a good measuring instrument is consistency. To test whether our subjects were reporting their Kinsey ratings in a consistent and dependable fashion, we used a basic principle of psychological testing: liars always make mistakes. Generally, several scales are used to measure the same characteristic, and the answers are compared. If there is a one-to-one correspondence between a score on the first scale and on each subsequent scale, the consistency is high. If there is no correlation between a score on one scale compared with the other scales, consistency is poor. The exact degree of consistency can be measured as a correlation coefficient, which ranges from zero percent for completely inconsistent to 100 percent for completely consistent.

The results of this analysis on our Kinsey scale data were encouraging. We calculated correlation coefficients for all six possible combinations of the four scale dimensions; for example, fantasy versus behavior would be one combination, attraction versus self-identification would be a second, and so on. All of the correlations were greater than 80 percent and the average coefficient was 92 percent. This indicates a high degree of internal consistency between the various facets of sexual orientation, a result that has been found by many sex researchers.

Reliability

The second requirement for a good measuring instrument is that it be reliable. Since our assessment of sexual orientation was based solely on self-report, how could we be sure the subjects were telling us the truth, or for that matter, that they even knew the truth about their sexual feelings? The only way we could truly measure a person's sexual thoughts would be by hooking him up to a mind-reading machine, which unfortunately does not exist. There is, however, a sexual lie detector called a penile plethysmograph. This instrument measures small involuntary changes in the volume of the penis, an indicator of sexual response. The subject's penis is fitted with a collar or sleeve connected to a sensitive gauge that records responses to potentially erotic stimuli. By comparing the responses to images of females and males, the direction of a person's sexual orientation can be inferred.

We did not have to use this procedure in our studies because other researchers using the penile plethysmograph have found a strong correlation between the penile responses and the self-reported Kinsey scores of the volunteers. These coefficients show that the Kinsey scores were not "just talk." They reflect erotic responses in the body itself.

Stability

The third qualification for a good measuring instrument is stability. The Kinsey Scale measures sexual orientation at one instant in time: the present. But to what degree do they reflect a person's sexual past and future? Is sexual orientation a constant feature of personality, or does it freely change back and forth?

To answer this question, Fritz Klein and colleagues developed a questionnaire that measures various dimensions of sexuality in the past, the present, and the ideal future. The purpose of developing the Klein Grid, as it's called, was to show that sexuality is a dynamic process, but when we used it we found a remarkable degree of stability. The straight men, on average, showed little change in their sexual self-identification, attraction, or fantasy. Only on the behavior scale was there any appreciable shift, mostly from occasional homosexual activities to exclusively heterosexual relationships. Most of these occasional same-sex contacts occurred during puberty and early adolescence, a period when boys do a lot of sexual experimenting.

The gay men also were quite stable in terms of sexual attraction and fantasy, but more variable for self-identification and behavior. This is not surprising given the considerable stigma attached to homosexual identification, especially for youngsters. A teenager or young adult is unlikely to call himself gay unless he's absolutely sure. Many of the gay men said their sexual contact with women was a result of social and psychological pressure to conform rather than any real interest or attraction. Some of the gay men found intercourse with women physically enjoyable, others did not, but most had tried it.

Brian and Michael Wilson's stories fit into this overall pattern of stability. Brian has been heterosexual all his life. As far as he recollected, the idea of having sex with another man never appealed or even occurred to him. One might argue that he has had such fantasies but concealed them from us, but he was quite frank about other sexual activities. By contrast, Michael had gay attractions and fantasies all his life. He identified, and to a certain extent, behaved heterosexually in his youth, but he said this was more the result of guilt and desire to conform than of any underlying attraction to women. One might propose that he is really bisexual and has suppressed his heterosexual attractions to conform with his gay peers, but this seems unlikely in light of the considerable anguish that his homosexual feelings caused him for many years.

This is not to say that all men maintain the same sexual orientation during their entire lives, nor that all changes in sexual direction occur early in life. There were a small number of exceptions to the rule of sexual stability. Consider, for example, the story of a man I know who we'll call Mark.

MARK: "I knew I was gay by the time I was 10. Some kid called me faggot. I didn't know what that meant, so I asked my mother. When she told me, I thought to myself, 'Yep, that's me.' Of course I didn't tell her that.

"I came out in college. In fact, I started our gay and lesbian students' alliance. Right from the beginning I was always more interested in having a steady boyfriend than in one-night stands. I met my lover, Barry, when I was 22 and he was 19, and we were together for almost seven years. I paid his tuition all the way through school. After he graduated he got a good job in another city, and we were only able to see each other on weekends. That was a big mistake. I remained faithful, but he started seeing other guys and eventually he dumped me for some hot young man. Two years later, Barry got AIDS. By that time, the new boyfriend was long gone, so I ended up taking care of Barry all through his illness. I was heartsick when he died. Despite everything he did to me, I still really loved him.

"Cindy came into my life just around the time Barry passed away. She was everything I'd ever sought in a lover: intelligent, caring, funny, supportive, you name it. And she was really good looking in a pixieish, perky sort of way. So we started sleeping together. At first that's all we did—sleep—but pretty soon it became a sexual relationship as well. Of course she knew all about my gay past, but she's a firm believer in bisexuality.

"Cindy and I got married seven years ago, and I haven't slept with a man since. I guess you could say I'm a gay man who loves a woman. If we ever broke up, I'm not sure if my next partner would be a man or a woman."

•

Here's another story that is both different and the same. It comes from Benjamin, the brother of two gay men.

BENJAMIN: "I was always attracted to women but also afraid of being rejected by them. Maybe it was my weight: I was a real fat boy in school. I also had to work all the way through high school, so I didn't have much time for dating. Then in my senior year I met a terrific girl who seemed to appreciate me for what I was rather than how I looked or money or any of that bull. I fell head over heels in love, and we were married right after I graduated. I have to admit, though, that I was always worried that she married me just to get out of her parents' house rather than any real love on her part.

"The marriage was okay for a while, but pretty soon it started falling apart. First she started nagging me: 'Why don't you make more money? Why can't we have a decent car?' and so on and so forth. Then she started just not being around. It was really lonely to come home to an empty house. Finally, she just walked out the door, and a week later her lawyer sent me papers claiming 'mental cruelty.' Can you believe it?

"That's when I started my gay phase. I guess the whole experience with my wife had really thrown me. I just didn't seem to have any desire to find another woman. Plus I had two gay brothers who were very supportive and had already taken me to a few gay bars. One night I went out to one of those bars by myself, and I was a big hit—fresh meat, I guess, or maybe it was my boyish good looks. Over the next two years, I slept with three or four different men, each of them for about six months. At first I called myself bisexual, but after a while it was easier to say I was gay. I should probably mention that I was more the pursued than the pursuer in those relationships.

"Then I met a new girl and everything changed. I'd actually known her through a church group for quite a while before we started dating, so we had friendship before romance. We went out together for almost three years before we got married. After my first experience I wanted to be really sure before making a commitment. Now we have four children, and our thirteenth wedding anniversary is coming up soon. I'm very satisfied in my marriage both emotionally and sexually. I don't regret my gay phase, but I certainly have no interest in returning to it, either. To me it was an important part of finding out what I really wanted in life."

The stories of Mark and Benjamin suggest that some people do change their sexual behavior in adulthood. But did they truly change their orientation? To us it seemed that both Benjamin's transitory gay phase and Mark's more long-lasting heterosexual period were more the result of rejection and emotional trauma than of any underlying shift in sexual feelings. It is also noteworthy that both men placed more importance on stable, committed relationships than on sexual adventure. Men who said they were more drawn to the purely physical, lusty aspects of sex seemed less likely to go through periods of heterosexual and homosexual behavior. Perhaps the most important point is that Mark and Benjamin represent the exception rather than the rule of stability. Most of the men we studied have always had the same sexual orientation and expect that it never will change.

The strongest evidence for the stability of sexual orientation is the consistent failure of attempts to change gay men to straight. Even some of the psychiatrists who devised the most Draconian of these "therapies," such as castration and electric shock, have admitted the futility of their efforts. The most famous example is Kurt Freund. During the 1950s, he developed psychological techniques to turn gay men into straight, and many of his patients told him they had indeed changed as a result of therapy. But Freund, unlike some reparative psychiatrists today, was a real scientist. He wanted to know if his patients truly had changed or were just trying to make him happy. To find out, he invented the penile plethysmograph. When he used it on men who said they had changed orientation, Freund found that most actually still responded sexually to men and not to women. Some had convinced themselves and even changed their behavior, but their underlying attractions were still toward men. There are a few modern examples of people who claim to have "cured" gays, but there is no hard scientific evidence.

Kinsey Scale Results

From these studies we concluded that sexual orientation, as we defined it, is a reasonably consistent, reliable, and stable measure of one aspect of sexuality. This allowed us to return to our original question: Is sexual orientation a discrete or continuous characteristic?

The answer is shown in fig. 1, which charts the distributions of the four Kinsey scale measurements for our volunteers. The striking feature is that almost all of the men were easily categorized as either gay or straight with few if any in between. In fact, there was not a single person who identified as a "three" on all four Kinsey scales, and only 3 percent of the men scored in the bisexual range of "two" to "four" on two or more scales. This is what would be expected for a discrete, or bimodally, distributed phenotype. If sexual orientation were a quantitative, or continuously, distributed phenotype, we would have found more of a bell-shaped curve. Brian and Michael Wilson were not the exception, they were the rule.

The largest separation between the two groups shown in fig. 1 is on the fantasy scale, on which more than 80 percent scored either a zero or a six, meaning definitely straight or definitely gay. The least dichotomous measure is behavior, but even on this scale, the overlap between the two groups represents only 6 percent of the sample. This suggests that sexual thoughts are more polarized than behavior. Since we were more interested in what goes on above the neck than below the belt, it seemed that our definition of sexual orientation as an "attraction" rather than an "action" was a good one.

COMING OUT

Determining that our subjects were gay, straight, or in between as adults wasn't enough. If we didn't know how, and most importantly when, they defined their sexual orientation, our results would be difficult to interpret.

When gays express their sexuality it is called "coming out of the closet," or "coming out." This is a process, not an event, whereby gay people define their sexuality, first to themselves and then to others.

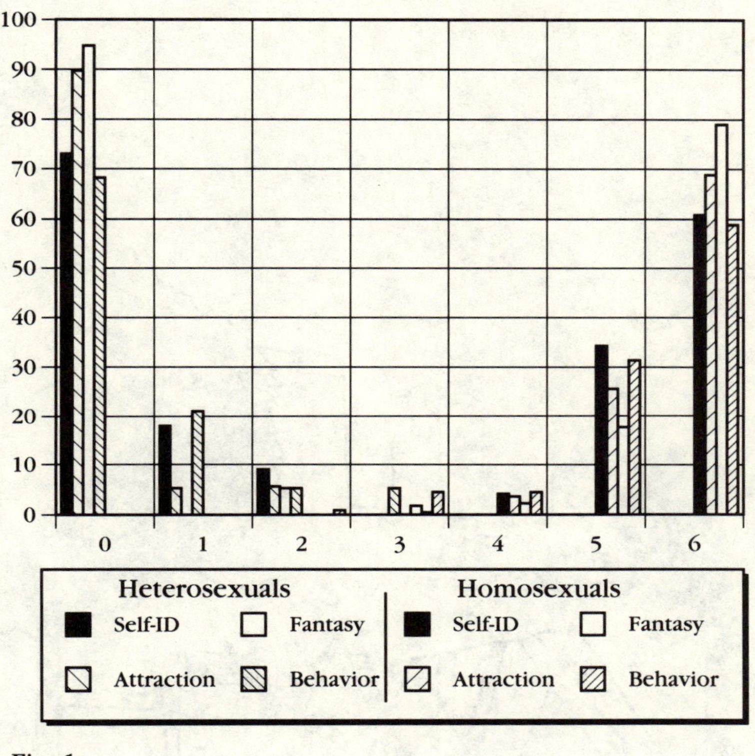

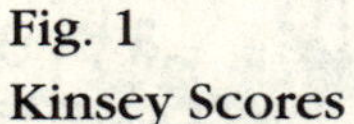

Fig. 1
Kinsey Scores

Gays come out in a diaspora: The vast majority of the people around them and of their role models—parents, relatives, friends, teachers, sports heroes, political figures, and fictional characters—are heterosexual. Moreover, most children get the signal early that being gay is "different," if not wrong or sick. So coming out is almost always a challenging process. One of our volunteers, a 32-year-old man who grew up in an upper-middle-class family and whom we will call Charlie, described his experience.

CHARLIE: "I always felt like I had to prove myself as a child, probably because other kids teased me about being a 'faggot.' The main way I

"This is Donovan. He's pretty sure he's gay, too."

Drawing by Crawford; © 1993
The New Yorker Magazine, Inc.

proved myself was through sports. Fortunately, I was better than average at just about all of them, especially tennis and swimming.

"My first crush was on a boy in about fourth or fifth grade, a classmate, but my first sexual contact was with a girl in sixth grade. I didn't do anything with a guy until I was in high school. I realized that I was gay, or at least bisexual, by junior year of high school. Of course that wasn't acceptable, since I knew I wasn't supposed to be that way. I was always popular and had a lot of girlfriends in high school, but I was also interested in guys.

"When I got to college I was still a virgin, and that bothered me. It was embarrassing. One night I got really drunk and stoned, and a woman dragged me into bed. That's how I lost my virginity. But I also had a big crush on a classmate from New York. We were very close freshman year. I went to his house for fall break and was waiting for him to attack me, but it didn't happen. In fact nothing sexual ever happened with him.

"During college I was fascinated with New York and the club scene. One weekend I went up to the city with some friends from college, and we went to Studio 54. I ended up going home with a man, a famous designer. The next morning I told my friends all about my adventure, except of course I changed the gender to female.

"I met my first true love, 'Phillipe,' in a philosophy class in Paris during my junior year abroad. He was the first person I told about being gay. Unfortunately, he already had a steady boyfriend, and after a while he stopped calling me and just blew me off. Of course I was devastated.

"When I got back from Paris I worked as a lifeguard for the summer and met a dancer whom I fell madly in love with. When I went back to college that fall I was very matter-of-fact about being with a man. I came out to pretty much everybody—roommates, friends, you name it. The reaction was pretty much, 'Well, that's okay,' but I didn't really want to talk about it much. That relationship lasted about nine months, until I felt like it was getting too serious and broke it off abruptly.

"I didn't come out to my parents until I was 23. My father didn't take it well, and things didn't start to change until my mom began going to PFLAG (Parents and Friends of Lesbians and Gays). Something really clicked. Now she's head of the local PFLAG chapter and has the PFLAG hotline coming right into the house. Once when I was home with my lover I heard her dispensing advice on how to find the lesbian bars in town. She's also on the board of the local gay and lesbian health clinic and is involved with the AIDS interfaith ministry at our church. Basically she's built a whole new life around gay activism.

"At first my dad would write checks for my mother's causes but didn't want to get personally involved. Then about three or four years ago, he really came into his own about what it's like to have a gay son. Now he sits on the board of an AIDS foundation and is very active at work."

When did Charlie "become" gay? Was it when he first felt a sense of being different from other boys in elementary school? Was it when he had his first same-sex fantasies or his first sexual experience with another male in high school? Was it at age 19 when he first told another person that he was gay or when he told his friends the next year? Or was it when he finally talked about his sexuality to his parents?

In terms of our genetic analysis, the critical question is when he first would have said he was gay to someone like me (an outside researcher). If I had interviewed Charlie in high school, I probably would have incorrectly categorized him as a nonhomosexual. An interview at age 30, however, would have caused me to rate him "definitely gay."

This is why all genetic analyses must look at when and how a trait is expressed. Suppose we were studying a trait, such as baldness, that usually is not expressed until midlife. If we studied only high school students, our analysis would be so distorted that we probably would end up "proving" that baldness is linked to cancer. That's because among the few bald high school students we would find, many would have lost their hair during chemotherapy. For the same reason, we wanted to be sure we were studying the sexual orientation of people who—if they ever were going to be gay—already were showing signs of being gay.

There was a second, more theoretical reason for being interested in the coming-out process. In trying to determine whether two alternative phenotypes, such as heterosexuality and homosexuality, have the same or different underlying mechanisms, it helps to compare how they develop. If both forms of the phenotype arise at similar periods in life, they *may* be due to alternative forms of the same pathway, which might be influenced by the same genes. If they arise at different times, they are less likely to be related. For example, it's obvious that having gray hair involves a different mechanism than having blond or brown hair, or straight or curly hair.

So for practical and theoretical reasons, we needed to know when homosexuality developed in our subjects and whether the process was similar or different for our gay and straight men. To find out, we used some of the following questions in interviews with all the subjects, heterosexual and homosexual, about their sexual awakening.

First Attraction: "Can you remember when you first had an attraction to a person that you now look back at and identify as sexual or affectional in nature? For example, a 'crush' or 'puppy love.' About when was that and who was it toward?"

Among the gay men, 96 percent had their first crush on a male, while among heterosexual men 100 percent were first attracted to a female. The ages at which gay men had their first attraction to men and heterosexual men had their first attraction to women were very similar, the median age being 10 years, and the range between 4 years and 16 years of age. The objects of affection were quite similar for both groups, although of opposite genders. The most popular choices were classmates, followed by teachers and actors and actresses. Pregay boys tended to be most attracted to stereotypical masculine heroes, such as Tarzan or Batman, while prestraight boys were inclined toward ultrafeminine types, such as Marilyn Monroe.

First Sexual Contact: "Can you remember when you first had sexual contact with another person that involved genitals? This could include 'fooling around' or 'playing doctor.' Please tell me when and with whom."

Among gay participants, 86 percent had their first sexual activity with a male, whereas 73 percent of the heterosexual men had their first contact with a female. The difference between the two groups is probably trivial: Most boys, whether they are pregay or prestraight, have more opportunity to "fool around" with other boys than with girls. For both groups there was a broad range of ages at which first sexual contact occurred (4 years to 31 years), with the median at about 12 years.

Puberty: "How old were you when you reached puberty: when you got pubic hair and could ejaculate?"

The median age of puberty was 12 years for both the homosexual and heterosexual men. Although other investigators have reported that the average age of puberty is slightly younger for gay men than for straight men, this was not apparent in our data. It is possible that our sample was too small and the interview question too broad to detect subtle differences. We did note, in agreement with others, that gay men who reached puberty later tended to have fewer sexual partners than men who reached puberty earlier.

Self-acknowledgment: "When did you first acknowledge your sexual orientation to yourself?"

When I tried this question on straight men, I got blank stares or "What do you mean?" Most heterosexual men have never really questioned their sexuality, and why should they? With the gay subjects, I got answers ranging all the way from 4 years old up to 30 years old. The median age was 16, with the greatest accumulation occurring between ages 11 and 19. For the subjects who self-identified early, a typical response was "I've always known I was gay, even before I knew the words for it." Among the men who didn't acknowledge their homosexuality to themselves until later in life, many reached puberty relatively late. Others were men who had sexual relations with women. Although some of the variability in the response to this question may reflect true developmental differences, disparities in how the subjects interpreted the question probably played an important role.

Acknowledgment to Others: "When did you acknowledge your sexual orientation to others, for example family members or people you don't really know, like me [the interviewer]?"

As with the previous question, this was designed for gay men. The age of acknowledgment ranged from 13 to more than 50 years of age, the median being 21. Most of the participants came out first to other gay people or to a psychologist or counselor, next to friends and family members, and last to outsiders, such as researchers.

Coming Out at 21

For genetic analysis, the key point about coming out is that the average age at which participants acknowledged their sexual orientation to outsiders is 21 years. This was substantially younger than the average age of the subjects: 36 years. Therefore, we assumed that most of our subjects who would be gay already showed signs of being gay.

On the mechanisms of sexual development, our research revealed two noteworthy features. First, the timing for homosexual and heterosexual men was quite similar. Both experienced their first sexual impulses—as manifested by crushes, fantasies, and sex play—at about the same ages. This is consistent with the theory that heterosexuality and homosexuality represent alternative forms of the same pathways, or different sides of the same coin. If this is true, it is possible that both outcomes are influenced by alternative forms of the same set of genes. Second, the gay participants' sexual direction was, in most cases, evident long before puberty, an observation well known to psychologists and psychiatrists, most of whom believe that sexual orientation is established within the first six years of life. This early manifestation implies that sexual orientation is a deeply ingrained component of a person's psychological makeup, which again is consistent with a genetic predisposition.

Chapter Four

BUILDING FAMILY TREES

Characteristics that are influenced by genes tend to be passed from one generation to the next. Since we were interested in the role of genes in sexual orientation, the first major goal of our project was to determine whether being gay runs in families. Researchers before us had described what it means to be gay and how people come out, but no one had looked at the patterns of sexual orientation in extended families.

If we found that the relatives of gay men had the same chance of being gay as anyone else in the population, we probably would have given up on the human sexuality project at that point. On the other hand, if we found that being gay did cluster in families, that would be an indication that genes might be at work. We knew from the start, however, that family trees alone could never prove that sexual orientation is inherited, because not everything that runs in families is genetic. If it were, there would be "Catholic genes" and "Presbyterian genes" to explain why those religions run in certain families, and there even would be "surname" genes that give members of a family the same last name. Obviously, family culture has a lot to do with certain family characteristics. So it would not be enough to show that homosexuality runs in families; it would be equally important to show that it is passed from one generation to the next through genes, not through family culture. The same would be true for our family studies of alcoholism and AIDS.

To understand how inherited characteristics are passed on in families, it's necessary to understand a few basic concepts of genetics: what genes are, how they are organized, and how they are transmitted from parents to their children.

GENES AND CHROMOSOMES

The root of all inheritance is the gene. In general, one gene encodes, or makes, one protein, which in turn catalyzes one biochemical reaction. Genes are so small they can't easily be divided, so they usually are passed on intact from parent to child. Genes are found on long, threadlike molecules called chromosomes, which are located within the nucleus of the cell. An average human chromosome, containing about 4,000 genes, is large enough that it can be seen with a light microscope.

Humans have two copies of each chromosome. One member of each pair is inherited from the mother, the other from the father. Each person has twenty-three pairs of chromosomes: twenty-two pairs called autosomes and one pair of sex chromosomes. The two members of each pair of autosomes are virtually identical; looking in the microscope it is impossible to tell which was derived from the mother and which from the father. The two sex chromosomes, known as "X" and "Y," are noticeably different from one another; the X chromosome is relatively large and long whereas the Y chromosome is small and dumpy. Men and women share almost the same genetic makeup, but males have an X and a Y sex chromosome, while females have two copies of the X. The sex of an individual is determined by the presence or absence of the Y chromosome, not by the number of X chromosomes.

Fig. 2 shows the typical array of chromosomes found in the blood cells of a male, in this case a gay man. The chromosomes have been arranged into a karyotype, or a portrait of the chromosomes in order of size. Almost every cell in the body contains the same set of forty-six chromosomes and thus the same genetic information, so each tiny cell contains all the information needed by the entire body. To make an analogy, each chromosome forms a single volume of a personalized set of encyclopedias containing everything necessary to be human. The genes are like the individual articles found in each encyclopedia.

Chromosomes from the male and female come together during

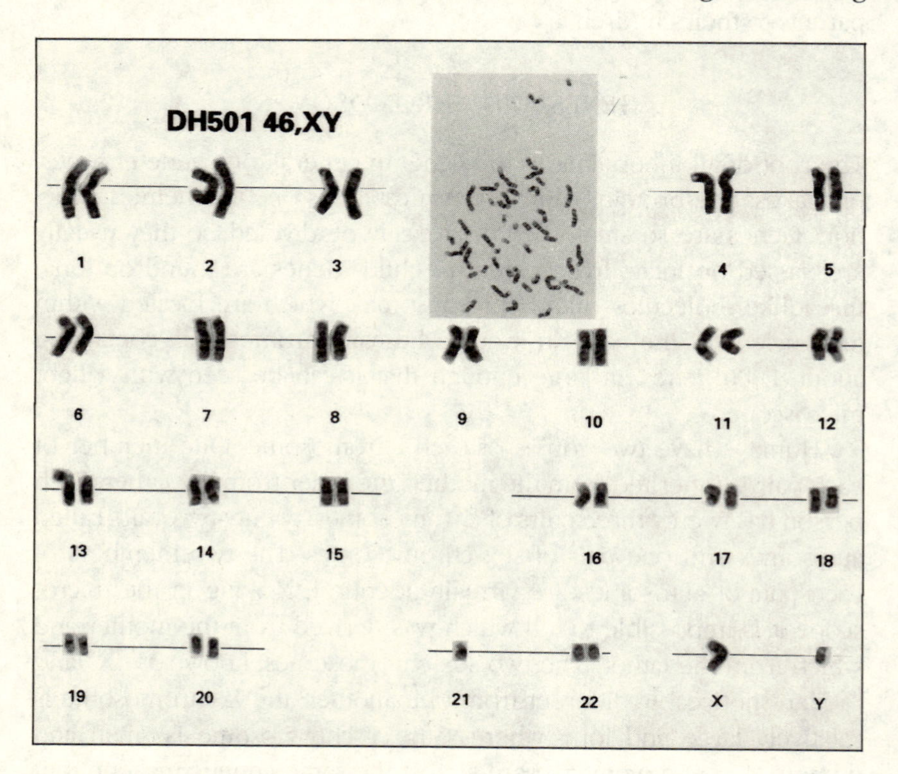

Fig. 2 A Male Karyotype

A typical male karyotype, or arrangement of chromosomes according to size, in this case of a gay man. The chromosomes are arranged in order of size. The inset shows how the chromosomes were arranged when the cell was opened.

conception, when a sperm cell fertilizes an ovum. Sperm and ova are special types of cells, called germ cells, that contain only one copy of each chromosome instead of two. Only one of each pair of chromosomes—either the one inherited from the mother or the one from the father—can end up in a germ cell, and the selection is determined purely by chance during a process called meiosis. Chance also dictates which of a large number of sperm cells will fuse with a particular ovum to form a fertilized egg cell. Once conception occurs and the male and female chromosomes have merged, all the cells in the new human will be almost exact copies of the original fertilized egg cell. Thus the genetic composition of a human depends on two throws of the dice, the first during the formation of the germ cells and the second at conception. These two actions shuffle the genetic information so thoroughly that a child can have her father's nose and her mother's eyes and other traits that seem to be a combination of both parents' physical characteristics or characteristics that are "gifts" from past generations.

PATTERNS OF INHERITANCE

Humans have known for ages that certain characteristics are passed on from parents to children. It wasn't until 1865, however, that an Austrian monk named Gregor Mendel discovered the basic laws of inheritance by studying the pea plants he bred in his spare time. And it wasn't until this century that people realized the laws discovered by Mendel are as true for humans as for peas.

The patterns of inheritance that Mendel observed in his pea plants were remarkably simple. The main reason was that, just by chance, all the traits he focused on were controlled by single genes. For example, if a plant contained one particular version, or allele, of a certain gene then it would produce smooth seeds; if it had a different form of the same gene then the seeds always would be wrinkled. Mendel's work was further simplified because all of the plants were raised in the same patch, thereby minimizing the effects of environmental variables, such as temperature, rain, and soil quality.

Most of the genetic traits, such as blood type, that first were studied in humans also are controlled by single genes. Even today, when scientists begin to study any human characteristic they think might be inherited—whether heart disease or sexual orientation—they always hope it will display the simple patterns that Mendel found in his pea plants. That's because it's much easier to track down the genes for such Mendelian traits than it is to analyze complex traits that may involve many different genes or a combination of genes and environment. So in early 1992, when I started interviewing gay men about their families, I hoped I might find a few "good" families with a large number of gays that fit a simple Mendelian pattern. The actual results turned out quite differently, but no less exciting.

Mendel's Laws

When I collected information about the families of the gay men I was interviewing, I recorded the results in the form of a family tree, or pedigree. I used squares to represent males and circles to represent females. Mates were connected by horizontal lines and parents and children by vertical lines. Definitely gay and lesbian individuals were indicated by filling in the square or circle, heterosexuals were represented by hollow symbols, and people I wasn't sure about were indicated by a system of dots, stripes, and other markings.

One pattern I was searching for in the pedigrees of our gay subjects was dominant autosomal inheritance. This would occur if being gay were caused by a single copy of one particular version of a gene on any one of the twenty-two pairs of autosomal chromosomes. If this dominant inheritance were the case, we would expect 50 percent of the siblings of a gay man, and at least one of the parents, to be gay also.

A second possible pattern was recessive autosomal inheritance. This would be the case if homosexuality occurred only in individuals with two copies of the hypothetical gay version of the gene for sexual orientation. In that case, I'd expect 25 percent of the gay man's siblings to be gay. Neither parent would have to be gay, because most likely each one would have a single copy of the "gay" form of the gene that would be offset by a copy of the "straight gene."

Somewhere in between autosomal dominant and recessive inheritance would be a semidominant pattern in which individuals with one copy of the gay version of a gene and one copy of the straight version would grow up with an intermediate, perhaps bisexual, orientation.

One afternoon at the HIV Clinic, after I'd finished an interview, the subject asked me why I was so interested in his family history. When I explained that we were trying to see if sexual orientation was influenced by genes, he made an interesting suggestion: "You should look at the sex chromosomes. That's what makes men different from women."

In fact I was interested in the sex chromosomes, not so much because of their role in making men and women different as because of their distinct patterns of inheritance. Specifically, fathers transmit their single Y chromosome to each of their sons and their single X chromosome to each of their daughters. Mothers transmit either of their two X chromosomes to both sons and daughters. This generates unique family trees and makes the sex chromosomes easy to track.

If being gay were caused by a gene on the Y chromosome, for example, homosexuality would be passed strictly from father to son. This would mean that all gay men would have gay fathers, and that gay men who had children would have only gay sons and heterosexual daughters.

A Y-chromosome gene wasn't likely, however, because obviously most gay men do not have gay fathers. A better possibility was a "gay gene" on the other sex chromosome, the X chromosome. If such a gene were transmitted through dominant X-linked inheritance, homosexuality would appear in both men and women carrying a single copy of the hypothetical "gay gene." If a father carried this version of the gene, all his daughters and none of his sons would express it. By contrast, if the mother carried the gay version of the gene, half of her daughters and half of her sons would be homosexual.

Any trait caused by dominant X-linked inheritance will be twice as common in women as in men in the population at large because women, who inherit two X chromosomes, are therefore twice as likely to have inherited the key gene than are men, who inherit only one X chromosome. Because other researchers had found that there are about twice as many gay men as lesbians in the population, this type of inheritance, at least in its strict form, didn't seem very likely.

A better scenario was a recessive rather than a dominant gene on the X chromosome, which would mean many more men than women with the trait. This is because recessive X-linked genes are expressed in women only when they have two copies of the variant gene. The chance of this happening is rare, so recessive X-linked traits usually are not expressed in women. By contrast, men will express the trait any time they carry the relevant allele, or variant, on their single X chromosome, because there is no second X chromosome to mask it. Another feature of recessive X-linked inheritance is that it would not require either parent of a gay man to be gay.

Two hypothetical examples of recessive X-linked inheritance are shown in fig. 3. In fig. 3A, the hypothetical gay form of the sexual orientation gene is originally present in the maternal grandfather, but it seems to disappear in his offspring. He has passed on his X-linked "gay gene" only to his daughters, who do not express the trait because they got a "straight" X chromosome from their mother. However, the

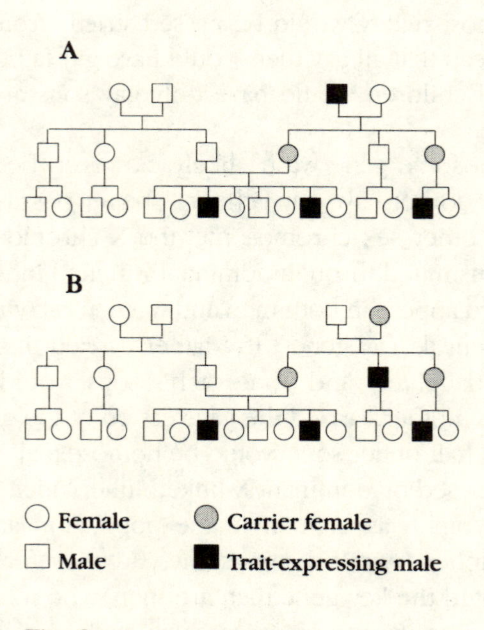

Fig. 3
Patterns of Recessive X-linked Inheritance

trait reappears in the male grandchildren in the next generation. They have inherited the gay variant from their mother, who was a silent carrier, meaning she had the gene but didn't express it. This is why X-linked traits are said to skip a generation.

Fig. 3B shows what happens when the "gay gene" is passed down from the maternal grandmother. In this case, homosexuality might appear in both the maternal uncle and cousin of the gay subject. The key feature of both diagrams is this: If sexual orientation is influenced by a recessive gene on the X chromosome, homosexuality will tend to be passed on to males through the mother's side of the family.

Complex Inheritance

After only a few weeks of interviews of gay men, it was clear that few families were going to show any of the simple patterns of Mendelian inheritance described above. This was not all that surprising. If sexual orientation were strictly determined by a single gene, it probably would have been discovered already. Besides, it is well established that most behavioral characteristics do not follow Mendel's rules, either because they are polygenic, meaning they are influenced by many different genes, or because they are multifactorial, which means they involve both genetic and nongenetic components such as the environment—nature *and* nurture.

As my collection of pedigrees grew, I began to consider some of the factors that could lead to more complex patterns of inheritance. The simplest departure from Mendelian inheritance occurs when the same gene can have many slightly different versions. If each combination of two different alleles, or versions, gives yet another variation, the family trees will show much more variety than those displaying Mendelian inheritance.

A different type of complication would occur if being gay could result from variations in multiple, independently acting genes. For example, if there were one "gay gene" that was autosomal dominant and another that was X-linked recessive, different families would show different patterns of inheritance. Moreover, if we examined their DNA, we would find that not all of these gay individuals would have variations in the same gene.

Multiple genes also could sway sexual orientation by interacting with one another rather than by acting alone. Suppose, for example, that being gay required variations in not one but two different autosomal dominant genes. A little mathematics shows that this would reduce the rate of homosexuality in siblings from one half to one quarter. Furthermore, neither parent would necessarily be gay. This could easily fool us into thinking that sexual orientation was a recessive trait. On top of that misleading clue, a DNA analysis of the family would show that not everyone who had the "gay gene" actually was gay.

We knew also that genes were only part of the answer. We assumed the environment also played a role in sexual orientation, as it does in most if not all behaviors. To most people, the environment means nonbiological factors, such as family upbringing, life experiences, and religion. To geneticists, however, the word "environment" means anything and everything that is not inherited, including some factors that are purely biological. So from our point of view, undergoing prenatal development in a womb swimming with male hormones is as much an environmental factor as growing up in a devoutly religious household.

Roughly speaking, an environmental factor can act with genes either independently or interactively. When it acts independently, an environmental factor by itself can cause a particular characteristic. If this hypothetical environmental factor were spread out randomly among the population, then many families might have only a single gay person, which could cause an underestimation of the genetic component of sexual orientation because homosexuality wouldn't appear to run in families. On the other hand, if the environmental factor were more common in some families than in others, it might mimic genetic inheritance and cause an overestimation of heritability. Such independent environmental factors would affect DNA studies in much the same way as multiple independent genes: Some individuals would be gay even if they didn't have the "gay gene."

An interactive environmental factor would cause homosexuality only in people who had the right genetic makeup. The effect of such interactive environmental factors would be similar to interactive genes: Few relatives of the gay subject would also be gay, leading to an underestimation of the genetic component of sexual orientation, and some individuals with the "gay gene" would be heterosexual because they lacked the necessary environment.

Given all the possible complications, I realized that interpreting the family trees of sexual orientation would not be simple. There was one thing, however, that worked in my favor: I was asking not just about immediate relatives, as was the case in the previous studies of twins and siblings, but also about aunts, uncles, cousins, and even more distant relatives. Because these relatives were raised in different households and by different parents, the effects of the environment provided by parents would be minimized. But to get this information, I needed a systematic way to collect family data and to verify the sexuality of the relatives.

THE FAMILIES

Once we had established the sexual orientation of a subject, we asked questions about the sexual orientation of the rest of his family. We explained that, "Depending on how well you know the person, and how open they are, you might know their sexual orientation for certain, you might have a hint but not be sure, or you might not have any idea. So please let me know how certain you are about your assessment and what it's based on."

We then asked the volunteers about the sexual orientation of all their adult brothers and sisters, nieces and nephews, aunts and uncles, cousins, their parents and grandparents. We didn't ask about relatives under 18 years old because too much guesswork is involved trying to characterize the sexual orientation of young people. I also asked about more distant relatives, such as second cousins or great uncles, who might be gay, lesbian, or bisexual. This information was recorded on a standard pedigree form, which is simply a chart of the family tree.

When I asked the subjects about their own sexuality, I used the structured sex history questions and the Kinsey Scale described previously. I knew the subjects were unlikely to have such detailed information about their relatives, so I devised a simpler rating system for family members. The original system I devised had four categories:

- Definitely gay: a Kinsey five or six, openly acknowledged to the person being interviewed or to other family members.
- Definitely bisexual: a Kinsey two to four as an adult, openly acknowledged.
- Possibly gay or bisexual: some reason to suspect something other than heterosexuality but not openly acknowledged.
- Heterosexual: a Kinsey zero or one, as indicated by an acknowledged attraction to the opposite sex.

Most of the gay men and their brothers we had interviewed personally fit into just two of these groups: definitely gay or heterosexual. We added the extra categories for the other relatives for two reasons. First, our family trees included women, who in general are more diverse in their sexual identity and expressions than are men, a fact that has been demonstrated in other studies and was confirmed in ours. Second, and most important, the subjects usually were less sure about their relatives' sexuality than about their own.

Not surprisingly, the most common category by far for relatives was "heterosexual." Typical descriptions were: "He's straight as a board," or "I sure hope she's heterosexual; she's got five kids."

The "definitely gay" category also was straightforward to score. For instance, one subject told us, "I know my brother is gay, because we talk about it all the time, we go out to gay bars together, and once we even dated the same guy." Another subject had this to say about his paternal aunt: "She went into the convent when she was 21, which Dad says is because she never liked guys. She got kicked out a few years later because she seduced a novice, and now they live together." Although we never met the two relatives described, we felt confident placing them in the "definitely gay" category.

There were very few men who were considered to be "definitely bisexual" as adults: only 3 out of 1,035 male relatives fit this category. Among female relatives, this category was somewhat more frequent: 4 out of 671 female relatives.

The most problematic category was "possibly gay or lesbian." Some of the stories made us almost certain that a relative was gay. For example, this was the description of a gay subject's maternal uncle: "I haven't seen him since I was in high school, way before I was out or anything. He came one Thanksgiving and brought along this young man—a real twinkie—and they were fawning over each other all during dinner. I had no idea what was going on, but later I remember my parents had a huge fight. After that we didn't see him anymore."

This is another description, in this case of a maternal great uncle: "The big scandal in our family was my grandmother's brother, a lifelong bachelor who everyone called 'The Poet.' He was in the foreign service, and when he came back from Egypt he brought along a houseboy who spent the rest of his life with him. This by itself was no big deal, but when he died and left all his money to 'Ali,' it caused quite a commotion."

Based on the descriptions, it's probable that both of these relatives were gay. But since they never acknowledged this, at least to the subjects or other relatives, we marked them down as "possibles" rather than "definites." Many other individuals were lumped into the "possibly gay or lesbian" category on even less certain grounds. For example, these are statements given by our subjects about relatives:

- "I think she was married once, but she got divorced almost right away. Anyway, she was a P.E. teacher, and you know what that means."
- "He never married, which was pretty unusual for a churchgoing Mormon."
- "She was in business way before that was considered 'normal' for a woman, and I think she was very close to her partner. But I could never imagine talking about sex to her. She was always so prim."
- "My cousin and I fooled around from when we were kids until I left for college. But now he's married and has a bunch of kids, so who knows?"

Who knows? That was the problem exactly. The subjects were too unsure and the evidence too thin, so we decided we needed confirmation by talking with the relatives themselves, or at least with other family members. Whenever possible, a face-to-face meeting was arranged; otherwise interviews were conducted over the telephone. The results were reassuring. Of the 69 male relatives identified as "definitely gay" and 4 female relatives considered "definitely lesbian," every single one agreed. Likewise, of the 28 men who initially were classified as heterosexual, 27 concurred and one declined to answer all the interview questions. Thirty-eight of the 39 women labeled heterosexual agreed, while one said she had a considerable history of same-sex activity.

Even after personal interviews, there were doubts about some people. Cousin Martin was one of the biggest question marks.

I learned about Martin through his cousins Roger and Peter, two openly gay brothers. I interviewed Roger first. He gave me the addresses of his brother, parents, and several aunts, uncles, and other relatives, but he asked me not to say anything to the rest of the family about his suspicions that Cousin Martin might be gay.

When I flew out to meet the family, the first interview was with Peter, Roger's younger brother. The two brothers were similar in appearance: neat preppies who liked rugby shirts and faded jeans. Both had been openly gay for several years.

After we finished with Peter's personal life, I took out a notebook and began to sketch his family tree, asking about the sexual orientation of everyone from parents to cousins. Peter listed his brother as definitely gay and only one other possibility in the family. Echoing his brother's comments, Peter said their Cousin Martin was "probably gay."

"He seems to like guys a lot more than girls, but way in the closet," Peter said, adding that his cousin had "looked funny" at him several times.

Next I met Cousin Martin's mother—the maternal aunt of Roger and Peter—who was delightfully spry for a woman in her seventies.

"I'll need to take a little blood," I said. "We can do that now or after the interview."

"Why don't we get that over so I don't have to worry about it," she suggested.

I had taken a week-long course at NIH on how to take blood, but this was one of my first interviews, and when she rolled up her sleeve,

I was confronted with a rookie blood-taker's nightmare: a fleshy arm like white jelly, pure of visible veins. I had a moment of panic that I'd blow the procedure and scare off the entire family. I imagined her calling me a quack and chasing me out of the house. Trying to appear confident, I swabbed on a little alcohol, gave her arm a few firm taps with my gloved fingertips, and struck. Mercifully, I hit a gusher on the first try. I soon filled six small vials with about eight teaspoons of blood. I took separate vials because I wanted to keep some blood as a permanent record, store some in our own freezer, use one vial for chromosome analysis, and have the rest to make DNA.

I started the interview with a few easy questions about her age, mailing address, and whether she favored her right hand or left. I asked about inherited diseases in the family and let her talk a while about her arthritis. I was slow to ask the sex questions because I always was a little worried about embarrassing older people. To put the subjects at ease, I tried to adopt the demeanor of a friendly insurance salesman; I wore a tie and carried my records and blood-collecting equipment in a black leather case that looked reassuringly like a doctor's bag. After a few minutes of pleasantries she told me she had been faithfully married to the same man for her entire adult life, which was enough evidence for me to place her in the heterosexual category.

She was proud of everyone in her family, and glad for the chance to talk about them, but not particularly helpful when it came to discussing sexuality. When I asked about the orientation of her nephews Roger and Peter, she replied that she'd heard that they were "that way" but couldn't really understand how that was possible. All that she had to say about the sexual orientation of her own children, including Martin, was that they were "fine."

Although in his 30s, Martin still lived with his mother, which immediately made me suspect he was indeed gay. At least until I saw him: He was big and beefy, crewcut and florid faced, with a potbelly that stretched a T-shirt out over a big leather belt holding up a pair of dusty jeans. I thanked him for agreeing to help with the study, and he crushed my hand in a working-man's grip. "Anything we can do to help," he said.

I asked questions about his childhood, and Martin appeared to be the stereotypical male: He had always liked sports, war games, and male activities. "Did you play with dolls or trucks?" I asked, following the standard format, but I already knew the answer. His first crush was on a girl, of course.

I didn't know what to think at that point, but I doubted he was gay. Still, each time he said he liked girls, I asked a question seeking more evidence. When he mentioned girls he liked in high school, I asked if he dated. Not much, he said. When he said he liked to sleep with women, I asked when was the last time. More than a year ago, he replied. All his fantasies and all his sexual experiences, he said, were with women. By any objective measure, he was a Kinsey zero: heterosexual.

There were bits and pieces of the story that didn't fit, though. He said he had never had a steady girlfriend or lover. The only women he'd slept with were prostitutes, and those experiences occurred when he was "whoring around" with his buddies in the army. Another thing that made me suspicious was that when I asked about masturbation, he denied doing it, referring to it dismissively as "kid stuff."

My greatest doubts were aroused by his answers to the standard questions about alcohol consumption. "I don't drink much, mostly beer," he said. No, he had never been in any trouble, such as fights or arrests, because of drinking. Nobody had ever complained about his drinking, and no, he'd never been to a meeting of Alcoholics Anonymous. He did concede that he would occasionally "tie one on."

Later when I spoke to Martin's sister Ellen, she told me that just the week before he had been arrested for drunk and disorderly behavior. She had picked him up, not for the first time, at the police station. She also told me that the courts had sent Martin to AA many times. She never once hinted that he might be gay, however.

Martin's aunt, Ilene, was the mother of Roger and Peter. She and her husband told me they had been worried about the boys since they came out as gay when they were in their 20s. The parents mostly were concerned about AIDS, but also that other people would reject the boys or be mean to them.

Because Ilene had two openly gay sons, she had thought a great deal about what "makes" people gay. But when I asked, "Do you think any of your other relatives might be gay, lesbian, or bisexual?" the answer was no. I didn't single out Martin or anyone else because that might have biased the response.

In the end we had the two gay members of the family pointing their fingers at their cousin but no one else confirming the suspicion. Martin did not tell the truth about one aspect of his life, his drinking, so he might have been lying about his sexuality as well—but I had no sure way of knowing. On the plane, heading back home, I wondered how to categorize Martin on the family tree. Was he gay, straight, or somewhere in between? The lesson for me was that personal interviews were not necessarily more informative about sexuality than family histories. If a person wanted to conceal his sexual orientation, he would conceal it from relatives and an outside researcher. I had to accept the fact that there always would be some uncertainty in our family trees—unless I started toting a penile plethysmograph in my doctor's bag.

By the time the plane landed in Washington, I'd made up my mind. I decided to condense the original four categories of sexuality for relatives into just two. "Homosexual" would consist of the definite gays only, and "nonhomosexual" would include everyone else. Admittedly this was a simplistic classification, but it had certain advantages. First, I felt confident the individuals rated as definitely gay really were gay. It seemed unlikely, given the stigma of homosexuality, that a heterosexual would masquerade as gay. Second, interviews were often impossible since the majority of the individuals believed to be questionable or possibly gay were from past generations and already had died. I reasoned it would be better to not classify them as gay than to guess. The final reason for not having a "possibly gay" category for people like Martin was that it seemed likely that gay men and lesbians would be unusually sensitive, even overly sensitive, to signs of "latent homosexuality" in their families.

After many trips, I eventually collected histories on 76 families. We hoped to see if the number of gays in these families was higher than the number of gays in the rest of the population, which would indicate homosexuality might be at least partly genetic. We also planned to search these families for alcoholism, mental disorders, and AIDS, to see if any patterns of inheritance were discernible.

During the first six months of interviews, I resisted the temptation to keep a running tally of the results. I didn't want to become falsely pessimistic because the rate of gays seemed too low, nor falsely optimistic if I hit a "jackpot" of gay relatives. It was hard not to take a peek, though, and by the time I had interviewed 76 of the subjects, I thought I might have enough information to make a reasonable, if early, judgment. All I needed was an excuse to count up the gays I had found in the families.

Chapter Five

A MOTHER'S LEGACY

I found the excuse I needed to take a first look at the data in June of 1992. The occasion was our weekly "show and tell" in the lab, where one person gets up and explains the current project or experiment. My turn was approaching, and I thought it was as good a time as any to prepare the first results from the family study. I stayed in the lab late one Friday, tucked away in the little corner that passes for an office. There is no window, only high shelves filled with books and papers, and two computers on a desk. It must have been around 8 P.M. because everyone else had gone home for the weekend. The only sounds were the constant humming of the lab equipment and freezers.

Looking through the computer files, I realized I had collected information on more than a thousand relatives of the 76 gay subjects. To check the rates of homosexuality in these relatives, I began by lumping together men and women from all the families and made a chart showing the broadest possible comparison. I decided to ignore nieces, nephews, and grandparents, fearing they might be too young or too old to give reliable information. This is how the data looked on the first run:

MALES AND FEMALES TOGETHER

RELATIVE	GAY/TOTAL	PERCENT GAY
Parents	1/152	0.6
Siblings	19/196	9.7
Aunts & uncles	11/397	2.8
Cousins	15/501	3.0

The results were not very exciting. I was sure that at least a few percent of the general population was gay, which meant that most of the families of our subjects actually included fewer gays than expected. The only hint of a higher rate was among siblings. Our gay subjects had 196 brothers and sisters, of whom 19 were definitely gay. Their rate of homosexuality of 9.7 percent seemed like a significant number but not startling. If we continued seeing such low rates for most of the relatives, it would be very hard to find any genetic link to sexual orientation.

I wasn't completely discouraged, however. I knew previous studies had suggested a considerable degree of difference for rates of homosexuality among men and women, so the next time I worked the numbers, I looked only at the male relatives of the gay male subjects:

RELATIVE	GAY/TOTAL	PERCENT GAY
Fathers	0/76	0.0
Brothers	14/104	13.5
Uncles	9/215	4.2
Cousins	12/243	4.9

This pushed the rate of gay brothers even higher, to 13.5 percent. Looking only at the men also raised the rates for gay uncles and cousins. Still, it was hardly a statistical breakthrough.

I checked the numbers on my charts again, just to make sure I had totaled the results correctly. For no real reason, I had drawn the charts with all the paternal relatives on the left and the maternal relatives to the right. I had put the total number of relatives in black and the "definitely gay" relatives in red. When I stepped back from the charts, I noticed there were many more red marks on the right side than on the left. Sensing there was some pattern, I ran the numbers again, but this time I separated the relatives on the mother's side and the father's side. The next chart revealed this:

RELATIVE	GAY/TOTAL	PERCENT GAY
Fathers	0/76	0.0
Brothers	14/104	13.5
Maternal uncles	**7/96**	**7.3**
Paternal uncles	2/119	1.7
Maternal cousins	6/103	5.8
Paternal cousins	6/140	4.3

This time we had something. There were many more gay uncles on the mother's side than on the father's side. Something was going on here. I wasn't sure what, but I knew we had found a pattern. I jumped up from my desk and ran into the hall, looking for someone to tell. The only person I could find was Juanita Eldridge, who despite having two teenage sons at home still put in long hours in the lab. She was busy trying to finish her work before leaving for the weekend. Even though she wasn't working on our project, her husband is a geneticist, and I knew she would be interested in what we were doing.

"You won't believe this," I said. "Look at this maternal loading."

To my dismay, she barely glanced at the data. She looked back at me and said, "That's right. Blame it on the mothers again." Then she walked briskly down the hall and back to her work.

I went home slightly dejected that such a neat set of figures could be reduced to a snide remark.

The next day was Saturday. I didn't plan on going in to the lab, but I was curious to understand what exactly we had found. I still wasn't sure what the pattern meant. I turned on my home computer and worked the numbers again. Why were there more gays on the mother's side? What could that mean—or did it mean anything?

One possibility that came to mind was "imprinting," a special type of genetic transmission, or inheritance, whereby a gene must pass through either the mother or the father—not both—to be active in the child. In other words, the gene would only work if handed down by the father in some cases and by the mother in others. Although the exact reason for this is still unclear, it seems to involve a sex-specific alteration of the structure of the gene. If the hypothetical "gay gene" could only be passed on by the mother, male homosexuality would tend to appear on the maternal side of the family; the father would be incapable of giving his sons an active form of the gene.

At the time, imprinting was in the news because it had been found to play a significant role in several important diseases. On my desk I happened to have an article about one of these, the fragile X mental retardation syndrome, which only is passed on by mothers. I remembered from the article that one of the hallmarks of imprinting is anticipation, a process whereby a trait becomes progressively more pronounced in successive generations. This probably is because the alterations in the structure of the gene are cumulative.

Perhaps, I thought, I should reexamine the family histories to see if homosexuality was becoming more "blatant" from one generation to the next. I wasn't sure how to measure something like that, but maybe the younger people were coming out earlier or had higher Kinsey scores than the older people. I quickly gave up on the idea, however. For one thing, I'd have to do longer genealogies, going back several generations, a kind of gay *Roots,* which would not be easy, considering the darkness of historical closets. Also, I already suspected from hearing so many coming-out stories that the age a person declares his homosexuality is far more a product of his social milieu than of any intrinsic factor.

Another explanation for the excess gays on the mother's side was simply that gay men were not passing on the gene, or any genes for that matter. Gay men, at least in present-day America, have considerably fewer children than do straight men; in our study, the gay subjects had ten times fewer offspring than their heterosexual brothers. Therefore, even if our subjects were gay because they had inherited a "gay gene," it was less likely to have come from their fathers than from their mothers. This would automatically result in more gay relatives on the maternal than on the paternal side, even if the gene was not on a sex chromosome.

The final possibility was the most interesting: recessive X-linked inheritance. Recessive X-linked traits always are expressed more frequently in men than in women, which could explain why there appear to be more gay men than lesbians in the population and why male and female homosexuality tend to cluster in different families. More importantly, X-linked traits always are passed to men through the mother's side of the family, which is the pattern we were seeing for homosexuality. Color blindness and hemophilia are well-known examples of such recessive X-linked inheritance.

I crossed my fingers and hoped we had an X-linked gene. If it were true, finding the gene itself would mean looking at the DNA sequences of only one of the twenty-three pairs of chromosomes, the sex chromosomes. But if the gene were on an autosome—not on the X or Y—we'd have to search through twenty-two pairs of chromosomes with no hint of where to start. It would be hard enough to search just one chromosome, let alone the entire human genome. The difference was like searching for an unknown article in only one volume of the genetic encyclopedia, or in twenty-two volumes.

I still didn't know if the family trees were enough to show X-chromosome linkage. Suddenly it hit me. The cousins were the key. If some of our subjects were gay because of an X-linked "gay gene," they must have inherited the gene from their mothers. That meant their maternal aunts also would have a chance of carrying the "gay gene" and passing it on to their sons, the maternal cousins of our main subjects. Maternal uncles also might have the "gay gene" and be gay, which was consistent with the data, but they could not pass it to their sons because there is no father-to-son transmission of the X chromosome.

Paternal aunts and uncles usually would not have the "gay gene" because they were on the wrong side of the family, which meant paternal cousins also were unlikely to be gay. So there was a simple prediction: A "gay gene" on the X chromosome would increase rates of homosexuality in the cousins through a maternal aunt but not in the cousins through a maternal uncle or a paternal aunt or uncle.

I looked at the family trees again, hoping I had been smart enough to record the sex of the blood-related parents of the 243 cousins of our gay men. No such luck. I knew I had the raw data at the lab, but it would take some time to compute the numbers. I went in to work, spread the family charts in front of me, and started pulling out all the cousins, then the parents of the cousins. A few hours later, I had the results:

RELATIVE	GAY/TOTAL	PERCENT GAY
Fathers	0/76	0.0
Brothers	**14/104**	**13.5**
Maternal uncles	**7/96**	**7.3**
Paternal uncles	2/119	1.7
Maternal cousins thru aunt	**4/52**	**7.7**
Maternal cousins thru uncle	2/51	3.9
Paternal cousins thru aunt	3/84	3.6
Paternal cousins thru uncle	3/56	5.4

That's it, I thought. I've got it. There clearly were more gay cousins through the sisters of the mothers. The only strange thing was there were as many maternal cousins as uncles who were homosexual. I had expected to find more gay uncles than cousins because, according to Mendel's laws, the uncles would share an average of 25 percent of their genes with our gay subjects, while the cousins would share only half as many, or 12.5 percent.

Then I realized that the explanation lay in the special rules that govern the inheritance of the X chromosome. Unlike on the other chromosomes, where uncles and nephews share twice as much genetic information as cousins, in the case of the X chromosome, maternally related cousins actually share more genes, an average of 37.5 percent. The explanation for this is that sisters share an average of 75 percent, rather than the usual 50 percent, of their X-chromosome DNA. That's because fathers have only one X chromosome to pass on, so girls get that one chromosome but may get either of the two X chromosomes from their mother. Finding as many or more gay cousins as uncles on the maternal side was another indication I should begin the gene search on the X chromosome.

I had been encouraged by the earlier numbers, but now I could barely sit still. I couldn't remember any time in my career where something had jumped up at me out of nowhere like this. I'd made my fair share of scientific findings during twenty years of working in labs, but in every case it was the result of long, hard slogging—of purifying proteins, running gels, reading sequences. Usually I had a pretty good idea of what I'd see next. But this data came out of the blue. It might sound a little hokey, but the only word that came to mind was "Eureka!"

SHOW AND TELL

I could hardly wait until the following Tuesday, when I was supposed to get up at lab "show and tell." The hour-long sessions are casual and informal, with lots of give-and-take; the idea is that brainstorming will make us all smarter. When the day came, I proudly herded everyone into a small conference room. The tradition is that the person who is presenting brings something to eat, but I was so excited about the pedigree results that I had forgotten to shop. Fortunately, Stella had a supply of Chinese junk food—dried peas and rice crackers—in her locker.

I stood up and passed out charts showing the rates I had found. I explained that past evidence of a genetic link to homosexuality was from studies of twins and nuclear families, but that no one had done a proper genetic study that included members of the extended family. There was an attempt in the 1940s, but it was hard to understand the family trees because the author described people as "musical" or "artistic." These might have been code words for gay, but it was difficult to tell.

I sauntered over to the chalkboard and quickly diagrammed how cousins actually have a greater chance of sharing a trait than uncles when the trait is linked to the X chromosome. These were the key rates, I explained, because they showed higher than normal numbers of homosexuals among family members raised in separate households. I was feeling pretty proud of myself at this point, but I cautioned that we still had a long way to go.

I noticed Stella Hu looking at me quizzically. "Excuse me, Dr. Hamer," she interrupted.

This was a bad sign. Normally Stella calls me Dean, and usually she doesn't interrupt.

"I read in the newspaper that 10 percent of men are supposed to be gay. I don't see why you're so excited about rates of 7 percent."

I was glad to hear that Stella was finally saying the g-word, but I also realized her criticism was on target. Without having a background rate of male homosexuality, without knowing what percentage of the general male population was gay, I had nothing with which to compare our data.

We talked for a while about how to find the background rate. My first impulse was to go through the published literature and use the agreed-upon number. But later, when I checked the research, I realized there was no such number; the estimates varied hugely. The 10 percent figure that Stella mentioned came from Alfred Kinsey's work in the 1940s. Kinsey himself never said 10 percent of the population was gay, but that 10 percent of the adult white males he surveyed—many of them prisoners—had been predominantly homosexual for a period of three or more years, sometime between age 16 through their 50s. Fewer than 4 percent of them had been predominantly or exclusively homosexual for all of their adult lives.

More recent random-population surveys in the United States, West Germany, and Canada had reported that anywhere from 0.3 percent to 6.2 percent of men identified themselves as gay. Estimates of homosexual behavior, as compared with self-identification, also had yielded a wide range of figures, from 1 percent for having had a homosexual partner during the previous year to 7 percent for any same-sex activity during adulthood.

When I looked closer at the jumble of surveys, I spotted variations in at least three factors that might throw off the results: the populations surveyed, the questions asked, and the definitions applied. Since our family study used a unique population group and our own specific definition of sexual orientation, it didn't make sense to rely on the published rates of homosexuality.

A second way of determining the background rate of homosexuality would have been to conduct our own survey by interviewing heterosexual men about the sexual orientation of their family members. The advantage would have been that I could demographically match the gay and straight index subjects and use the same questions and definitions for both groups. But there was a potential problem with this approach. Heterosexual men probably would not be as aware of homosexuality in their relatives as gay men, and therefore I was likely to underestimate the population rate of gays.

The best solution would have been to use gay men who were adopted and study their new, nongenetic families. That would neatly separate the questions of environment and genetics. If homosexuality were a result of family environment, then the new relatives of an adopted gay man should have the same rate of homosexuality as the blood relatives of gay men. It would be costly and time consuming to find a significantly large number of adopted gay men, but it's a study that ought to be conducted at some point.

One day during the time I was trying to come up with a background rate, I ran into Elliot Gershon from NIMH. Elliot had made his early reputation by conducting careful, large-scale studies of the incidence and familiarity of various psychiatric disorders. When I explained my difficulty coming up with a background rate, he smiled and said, "I know the answer." He had asked all the participants in his epidemiological surveys about same-sex activities, but he had never reported the results because homosexuality was no longer considered a psychiatric diagnosis. The information still existed, however, and he promised to dig up the data from the archived computer tapes.

The following week I got help from another source. My colleague Angela Pattatucci was up for lab "show and tell" and talked about the preliminary results of her survey of lesbian families. She had some interesting findings but was confronted by the same problem I had: She didn't have good baseline figures with which to compare her results. After the seminar she asked if I had any numbers on lesbian relatives. That triggered my memory. I had collected information about female relatives but then discarded it to concentrate on the men. Why couldn't I use my data on the families of the gay male subjects to estimate the background rate of female homosexuality for Angela? Then I could use her data on the families of lesbians to estimate the background rate of male homosexuality.

This trade would work, we decided, because other researchers had shown that male and female homosexuality were largely independent and usually didn't show up in the same families. Even if there were some overlap, the worst that could happen was that we'd overestimate the background rates and therefore underestimate the degree to which homosexuality was clustered in families. That would lessen, not increase, the chance of finding a genetic connection.

There were clear advantages to this approach. For one, Angela and I were using the same interview format and questions. In fact, when there were scheduling conflicts we sometimes interviewed each other's subjects. More importantly, we were using the same stringent definition—that only people who openly identified themselves as gay or lesbian would be categorized as "definitely homosexual." We also were confident, from the many interviews we had conducted, that lesbians and gay men were equally aware of homosexuality in both male and female relatives. This would eliminate the reporting bias that might have come from comparing the family histories of homosexual to heterosexual index subjects. Best of all, we already had the necessary data in our notebooks and computers.

Within a short time, I had prepared the information from Angela that I needed for a background rate. In this case the primary subjects were the lesbians, but we were looking at their male relatives:

RELATIVE	**GAY/TOTAL**	**PERCENT GAY**
Fathers	3/150	2.0
Brothers	8/169	4.7
Maternal uncles	2/152	1.3
Paternal uncles	3/94	3.2
Maternal cousins thru aunt	2/124	1.6
Maternal cousins thru uncle	1/106	0.9
Paternal cousins thru aunt	3/95	3.2
Paternal cousins thru uncle	3/94	3.2
Nephews	0/24	0.0
Grandfathers	1/236	0.4

Adding it all up, 26 out of the 1,244 male relatives of the lesbians were gay, giving a background rate for homosexuality of 2.1 percent. When I eliminated the nephews and grandfathers from the data, reasoning that they might be too young or too old to give reliable results, the corresponding figures were 25 gay out of 984, or 2.5 percent. Removing the fathers, who were less likely than average to be gay, because they had children, changed the figure only slightly to 22 gay out of 834, or 2.6 percent. In the final calculation I also left out brothers because I thought they might have an increased chance of being gay because of the same family environment. This gave me 14 out of 717 uncles and cousins who were gay, corresponding to a 2.0 percent population incidence of male homosexuality.

Then I went back to my data on the gay men and their relatives to compare it with our new estimates of the background rate of male homosexuality. To be extra careful, I actually made three separate calculations; the first used the lowest background rate of 2.0 percent, the second used the higher background rate of 2.6 percent, and the third made a direct comparison between the relatives of the gay men and of the lesbians for each type of family member.

The statistical analysis confirmed my suspicions. Of the eight types of male relatives considered, only three had significantly elevated rates of homosexuality: brothers, maternal uncles, and maternal cousins through an aunt. All of the other rates were not significantly different from the background rate. Moreover, it made little difference which background rate I used. These results quantitatively confirmed the maternal skewing that I had sensed the first time I jumped up and showed the data to the skeptical Juanita Eldridge.

One point should be emphasized about the background rates used in our study. They are not, and never were, intended to be accurate measures of the incidence of same-sex behavior or orientation in the population at large. Our only purpose was to determine the appropriate background rate for our family study of gay men, and for that it was essential to apply the same stringent definition of male homosexuality to a comparable population. I think we made a good choice by using the lesbian families, at least for this limited purpose. In truth, I don't think there is such a thing as "the" rate of homosexuality in the population at large. It all depends on the definition, how it's measured, and who is measured.

A few weeks after I'd completed the pedigree analysis, I received a thick envelope from Elliot Gershon with the information he had promised. Not only had he retrieved his old psychiatric data, he'd written a paper about it. Elliot's measure of sexual orientation was not very sophisticated; it consisted of the single question, "Have you ever had sex since age 18 with a person of the same sex?" The rates of people answering yes were 3 out of 290 (1.0 percent) for those with no psychiatric diagnosis, 11 out of 532 (2.1 percent) for individuals with no major diagnoses, and 28 out of 725 (3.9 percent) for the total population studied, many of whom had schizophrenia or other mental illnesses that might, in Elliot's opinion, have caused them to be sexually "disinhibited."

Although these figures were collected in a different way and with a different intent than our own figures, they nicely bracketed our background rate of 2 percent. Since I suspected that somewhere down the road we'd be challenged about our "low" figure for the population incidence of male homosexuality, it was comforting to have supporting data from someone I trusted.

CONTROLS, CONTROLS, CONTROLS

Just when things were starting to fall into place, I received a phone call that made my head spin. When I started the new project I installed a separate telephone line and answering machine for my office so volunteers could leave messages in private, without having to go through someone in the lab. One day shortly after I'd started the family study, I received a message from Tony, an HIV Clinic outpatient scheduled to be interviewed that afternoon. "I'm sorry, Dr. Hamer, but I have to cancel my appointment this afternoon," he said. "Anyway, I don't have any gay relatives, so you probably wouldn't be interested in me."

Damn, I thought. How did that happen? Even though I'd been careful not to mention anything about families or genetics in the flyer that was passed out at the HIV Clinic, it was obvious that Tony knew what I was up to, probably from talking with another participant. If Tony knew, what about the others? Might our results be compromised by a bias toward volunteers who thought they'd be "interesting" to us because they had gay relatives?

Fortunately, there was a simple way to test this. The participants from the HIV clinic, Whitman-Walker, and the Triangle Club represented only a portion of the target populations. In other words, we got only the people who volunteered from those places. But with the group organized by gay and lesbian Christian Scientists, Emergence, I was able to interview every single member, so there was no room for volunteer bias. Once I'd collected all the pedigrees, it was a straightforward matter to compare the results for the Emergence group with the others. The analysis showed no significant differences in the rates of homosexuality in the relatives of the Emergence subjects versus the relatives of the other volunteers. This made me more confident that our pedigree results were not tainted by volunteer bias; Tony's call was a fluke.

It wasn't long after I had dismissed that possible bias that Juanita Eldridge, my colleague who had claimed I was unfairly blaming mothers, came up with another possible bias. "All you've shown," she insisted, "is that children are closer to their mothers and her relatives than to their fathers. Any mother could have told you that."

I hadn't thought of that. Might the reported excess of gay male relatives on the maternal side be an error that occurred because the participants simply knew more about their maternal relatives than the paternal ones?

We came up with two ways to check this. The first was to look at the distribution of lesbian relatives of the gay male subjects. If it were true that the subjects were better informed about the sexual orientation of their maternal than paternal relatives, we should have found more lesbians on the maternal side than on the paternal side. But this was not the case. There actually were slightly more (but not significantly more) lesbians who were related to the subjects through fathers than through mothers.

A second control was to look at the gay male relatives of the lesbian subjects being interviewed by Angela Pattatucci. Again there was no significant difference between the rates of homosexuality in the exclusively maternal versus paternal branches of the family. These results show that there really were more gay men—not just *known* gay men—on the maternal side than on the paternal side of the gay male subjects' families.

THE EVIDENCE FOR GENES

The pedigree study failed to produce what we originally hoped to find: simple Mendelian inheritance. In fact, we never found a single family in which homosexuality was distributed in the obvious sort of pattern that Mendel observed in his pea plants. But in other ways the results were encouraging. First, we detected significantly elevated rates of homosexuality even in relatives who were raised by different parents in different households. This favored an explanation based on genes rather than the rearing environment. Second, we quite unexpectedly found more gay relatives on the maternal than on the paternal side of the family. This immediately suggested a specific mode of inheritance: X-chromosome linkage. The ultimate test of this idea would involve looking at the DNA itself. But to perform that experiment, we would need DNA from a specific type of family: those with two gay brothers.

Chapter Six

LOOKING FOR LINKAGE

One evening around 6 P.M. I was carrying some blood samples to the storage freezer when I sighted a familiar figure ambling down the hall toward the elevator. It was David Botstein—the once enfant terrible, now rapidly becoming éminence grise, of modern molecular genetics. It was Botstein who had, in 1980, proposed that it might be possible to map the entire human genome by using random bits of variable DNA called markers. And it was Botstein who later showed that it was possible, at least in principle, to use a dense map of such markers to identify genes even for complex, non-Mendelian traits. Although Botstein himself had never actually isolated any markers or mapped any human genes, he was widely regarded as one of the most important and critical thinkers in the field.

I was interested to get Botstein's reaction to our project, so I gave a holler and trotted down the hall to intercept him at the elevator doors. I had known him for years, but he was in a hurry and didn't have much time to talk. He said he had been visiting NIH to give a seminar and now he was leaving.

Quickly I gave him the rundown of what we had been doing and told him about the family survey and our hunch that there was a genetic factor in homosexuality that was inherited from the mother.

He cut me off abruptly, even before I had finished telling him about the X chromosome. "You should do linkage analysis on pairs of brothers," he said. "That's the only evidence that I'd believe. That's how we're working on manic depression."

With that he trundled into the waiting elevator and disappeared.

Botstein had dropped an interesting tidbit about his own work that helped convince me we were on the right track. If he was looking for a genetic link in a behavioral trait as complex as manic depression, then we weren't crazy to be looking at homosexuality. And he was doing linkage analysis on siblings, which was the direction we were taking, too.

Our family studies had shown that male homosexuality runs in families in a way that could mean genetic transmission, but we were still a long way from proving that genes were involved. There were too many other possible scenarios, such as a type of family environment that was replicated from one generation to the next. Similarly, finding more gays on the mother's side hinted at a gene on the X chromosome, but there were many alternative explanations.

Even if genes were involved, it certainly didn't look like a case where one gene explained everything and was neatly transmitted according to Mendel's laws. If homosexuality were the result of such a single gene, then half of the brothers of our gay subjects also should have been gay, since they shared 50 percent of their genetic information. Twenty-five percent of the uncles should have been gay, and 37.5 percent of the cousins also should have been gay, assuming the X chromosome was involved. Remember that X-linked traits have their own rules. If a color-blind man, for example, has three daughters, each one will carry the gene but won't be color blind. Each son of each daughter has a 50 percent chance of inheriting the gene and being color blind. Our rates for homosexuality should have shown the same kind of pattern. Instead, the rates we had were higher than for the rest of the population but nowhere near these Mendelian levels.

FAMILY ENRICHMENT

One possible explanation for the rates of homosexuality we were finding was that in some families genes for homosexuality were more important than in others. In other words, some people were gay partly because of their genes and some were gay for other reasons. The way to isolate the genetic component was to enrich for, or select, families that appeared to have the gene. Family enrichment is a deliberately biased technique, which is what makes it effective. Imagine that family enrichment is like a powerful lens. Choosing families likely to have the gene is the same as raising the power of the lens: It helps you see more clearly, but if nothing is there, you still won't see anything.

The basic idea of family enrichment, also known as genetic loading, is simple. Genetically influenced traits run in certain families because the relevant genes are clustered in those families. It follows that the genes of interest are "enriched" in those families where the trait is prevalent. Hence those are the best families in which to search for the gene. For example, if we were studying the type of breast cancer linked to a specific gene, it would be most productive to look for the gene in families with high rates of breast cancer.

One indication that a family is enriched for a gene is a pair of siblings who express the trait. In our case, this meant finding pairs of gay brothers. In a simplified illustration, assume that being gay could be caused either by a gene or an environmental factor. Assume also that the gene is on the X chromosome, that it's operative only in men, and that 2 percent of the general population is gay. Now suppose the gene and the environmental factor are equally prevalent. In that case, 50 percent of a random sample of gay men should have the "gay gene," but 50 percent will not. On the other hand, the laws of mathematics show that in a group of gay brothers, 97 percent will carry the "gay gene" and only 3 percent will lack it. Thus the gene has been enriched by a factor of two. A similar calculation shows that a gene found in only 5 percent of a random selection of gay men would appear in 55 percent of gay brothers, an enrichment of more than tenfold.

So pairs of gay brothers served as signposts that their families were likely to have the gene for homosexuality, if such a gene existed. The other reason for choosing brothers was that siblings are the smallest family unit that can easily be used for genetic analysis. When looking at DNA, it is far easier to compare and contrast related people than unrelated individuals. The closer the relation, the easier the comparison.

Our family studies showed us what kinds of families our gay brothers should have to be most useful in the genetic analysis. First, the family should have exactly two gay brothers. If there were only one gay man there'd be no enrichment for the gene, and if there were more than two, we ran the risk of selecting rare or unusual genes. Second, there should be at most one lesbian in the family. This is because the family studies showed that male and female homosexuality were not commonly found together, and we wanted to use "typical" families. Finally, we did not want families with gay fathers and gay sons, because this pattern also would not be consistent with X-chromosome linkage.

Once we determined that our families had two gay brothers and met the other criteria, then we would find out about the sexual orientation of the other male relatives. If our theory about a genetic link to homosexuality were correct, then the gay brothers should have a higher than average rate of gay maternal uncles and cousins. If the theory were not correct, then we could analyze all the families with gay brothers in the world and still not find higher rates of homosexuality among the other male relatives. The first step was to find the gay brothers.

The "Santa Claus" Brothers

When I met Donald and Erik Mullen in the reception area of the NIH Genetics Clinic, my first thought was that they would make perfect department-store Santas. Donald was 65 and had a full head of pure white hair, a bushy white beard, and an emerging potbelly. Erik, at 68, had the same white hair, a more restrained white beard, and an ample "bowl full of jelly." Both were cheery and chatty. Most interesting to me, both were gay.

Donald Mullen heard about our study through the mailing sent through PFLAG, the gay and lesbian family organization. He sent me a note saying he had a gay brother and that they'd be willing to participate if the NIH would pay their travel expenses. I quickly agreed.

Donald had identified himself as gay for more than thirty years—almost as long as most of our other volunteers had been alive. As a child, Donald was a loner who felt "different" from other boys and was teased as a sissy because he preferred to play school instead of sports. His first crush, at age 7, was on a male piano teacher, and his first sexual contact was at age 12, with the boy next door. He had little interest in dating or socializing during high school. His first adult sexual experiences were in the navy, when he went to a bar on liberty, had a few drinks, and was picked up by a man.

After leaving the navy, Donald took a job as a teacher and fell in love with one of his colleagues. When their relationship was discovered, both of them were promptly fired. Donald moved to California, determined he was not going to be gay anymore. Instead, he fell for a younger man, and this time the consequences were more severe. When the young man told his parents about the relationship, they had Donald arrested on charges of sodomy. The judge gave Donald an option: jail or a psychiatric hospital. Donald chose the hospital.

He spent two years there. The psychiatrists must have considered him a desperate case, because they decided to perform a brain operation that involved drilling several holes in his skull. Although Donald still is unclear about the exact purpose and details of the operation, the immediate result was obvious: He was paralyzed on the right side of his body for several months. Donald told me that another man in his therapy group committed suicide.

After Donald's release from the hospital, he made a determined effort to go straight and was married a year later. Although his wife knew about his history, she felt she could help him change. The relationship lasted for six years, and they adopted two children. But Donald still was strongly attracted to men, and by his early 30s had concluded that he "couldn't change." Eventually his desires for emotional and sexual gratification overcame his hope for a stable family life, and he obtained a divorce.

Ever since, Donald has lived an exclusively gay life. Although he's had several long-term relationships with men, he currently lives alone. He takes great pride in his work as a counselor, and when he visited the NIH, he brought an article about combating homophobia in the counseling profession.

Erik Mullen, like his brother, was teased as a child because he didn't like sports and preferred to spend his time painting and modeling clay. He remembered having his first attraction to a boy at age 7 and his first sexual experience at around 12. Also like his brother, his first adult sexual experience was in the armed services, in this case with a fellow soldier who became his lover throughout the course of World War II in the Pacific theater.

Although Erik never doubted he was homosexual, he hid it as best he could and married at age 22. The marriage lasted thirty years and produced one child, a daughter. Sexual relationships with his wife were rare, and she sometimes mocked him because of his lack of interest. Erik said that at age 48 he no longer could stand the pressure of living a lie and attempted suicide. This led to psychotherapy and several unsuccessful attempts to "cure" him. Only at age 52 did Erik get divorced and start his new life as a gay man. He recently retired from his job as an art teacher and currently has a lover.

Donald and Erik were not the only gay men in the Mullen family. They had a nephew, the son of one of their two sisters, who they said was homosexual and had died of AIDS. They also had an uncle, the brother of their mother, who Donald claimed had sodomized forty-nine men in the farming community where he grew up. (How Donald came up with such precise data, I'm not sure.) Their family tree displayed the pattern that I had come to expect from the family survey: All four gay men were related through the maternal line.

The "Santa Claus" brothers, as I came to think of them, were fascinating on many levels. They were remarkably similar in appearance, personality, and early sexual experience. They shared an immense trauma about being gay, and both came out of the closet at a relatively late age. Their family history was the most interesting of all. I strongly suspected that if there were indeed "gay genes," families like this one would be the right place to look.

By the spring of 1993, I had found 40 families with gay brothers, including the Mullens, that fit the requirements. Two of these were from the initial survey, and the remaining 38 were recruited through advertisements. The main thing I wanted from these families was their DNA, but I also wanted to test the theory of family enrichment. If the theory of a genetic link on the mother's side was correct, these families should show the same pattern observed in the initial survey; that is, an excess of gay maternal uncles and maternal cousins through an aunt.

Once again, I sat down at the computer to calculate rates. My task was simplified by the fact that I'd already separated the cousins into the four possibilities of maternal or paternal and through an aunt or an uncle. Within an hour or so I'd generated the following table:

	Random Survey		*Families with Gay Brothers*	
RELATIVE	**GAY/TOTAL**	**PERCENT**	**GAY/TOTAL**	**PERCENT**
Maternal uncle	7/96	7.3**	6/58	10.3***
Paternal uncle	2/119	1.7	1/66	1.5
Maternal cousin thru aunt	4/52	7.7**	8/62	12.9***
Maternal cousin thru uncle	2/51	3.9	0/43	0.0
Paternal cousin thru aunt	3/84	3.6	0/69	0.0
Paternal cousin thru uncle	3/56	5.4	5/93	5.4

** $P < 0.01$ compared to estimated population frequency of 2.0 percent
*** $P < 0.001$ compared to estimated population frequency of 2.0 percent
P is a measure of the statistical confidence level.

These results confirmed my hunch. Just as in the initial survey, the highest rates of homosexuality were found in maternal uncles and cousins through an aunt. Moreover, the significance of the results was upgraded from "two stars" ($P < 0.01$) to "three stars" ($P < 0.001$).

Given that the two surveys were conducted on independent groups of gay men—one picked more or less at random and the other recruited by advertising—it seemed increasingly unlikely that this maternal effect was just the luck of the draw.

The idea of family enrichment also was borne out, albeit less obviously. In the case of the maternal uncles, rates were increased from 7.3 percent to 10.3 percent. For cousins on the mother's side related through an aunt, the rates were increased from 7.7 percent to 12.9 percent. Although these increases were not statistically significant, they certainly were in the right direction.

LINKAGE ANALYSIS

Family studies alone can never prove a characteristic is caused by a gene, especially when the trait doesn't obey Mendel's laws. Even if family studies could prove such a thing, they wouldn't say anything about what the genes actually do—what they code for, how they act in the brain, or how they differ in different people. The only way to prove that genes are important, and ultimately to understand how they act, is to look directly at the genetic information itself. The technique used to do this is called linkage analysis.

Mendel was a lucky man. When he crossed two types of pea plants —a tall one with wrinkled seeds and a short one with smooth seeds— he found that the baby pea plants had equal chances of all four possible outcomes: tall and wrinkled, tall and smooth, short and wrinkled, and short and smooth. The reason for this mixing and matching of traits was that Mendel, just by chance, was working with genes found on different chromosomes.

Half a century later, an American geneticist named Thomas Morgan made an observation that seemed to contradict Mendel. Instead of pea plants, Morgan worked with two kinds of fruit flies: purple-eyed, black-bodied mothers and red-eyed, white-bodied fathers. In this case, the traits never mixed and matched. The babies always looked like one of the parents, not a combination; they always had either purple eyes and black bodies like the mother, or red eyes and white bodies like the father.

What Morgan had discovered is that the genes for eye color and body color in fruit flies are "linked." Genes for both traits are located on the same piece of DNA on the same chromosome, and since pieces of DNA usually are passed whole from one generation to the next, eye color and body color always are inherited together. When traits are linked, one can be used to help track the other. Finding enough links enables geneticists to draw maps showing where genes are located.

Linkage analysis in humans is based on two fundamental principles of genetics. First, if a gene influences a trait, then individuals who share the same form of the gene have a good chance of sharing the same form of the trait. This applies to unrelated people, but the influence of a particular gene usually is most obvious in close relatives, such as brothers. That is because in some cases the gene will only work with other genes or in a particular environment, conditions shared more often by close relatives. In other words, if a gene must interact with another gene or only expresses itself in a certain environment, the person most likely to share the second gene and the environment is a close relative, such as a sibling.

The second principle that influences linkage is that genes found close to one another on a chromosome usually are inherited together. Because of simple chemistry, two genes that are "close together" are, by definition, found on the same DNA molecule. DNA is tough and wiry; only rarely do its long strands break in two. As a result, genes that are located on the same piece of DNA almost always travel together into the germ cells that make up the fertilized egg.

From these two principles emerges a third. If a trait really is influenced by a gene, then people who share the trait have a good chance of sharing a piece of DNA close to the gene. Conversely, if no gene exists, then there is unlikely to be a shared piece of DNA. This means linkage can be used not only to map genes but also to determine whether or not a gene even exists.

A good way to understand this is with the simple example of color blindness, a trait that follows Mendel's laws. Color blindness is controlled by a recessive gene located on the X chromosome. In fig. 4, the black squares represent color-blind men, and the white squares and circles represent men and women who can distinguish colors.

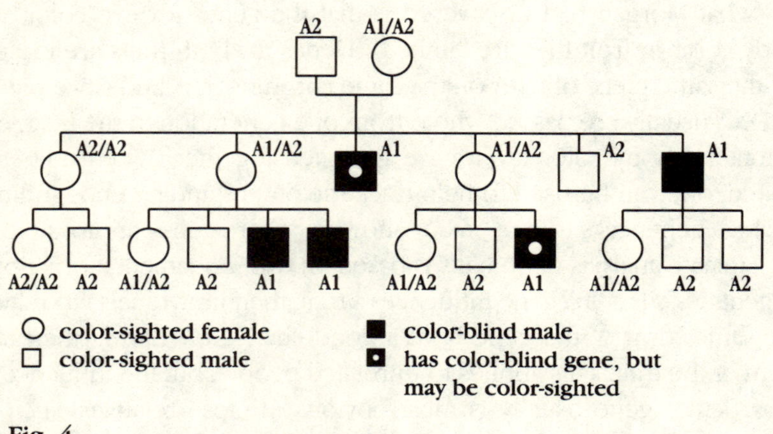

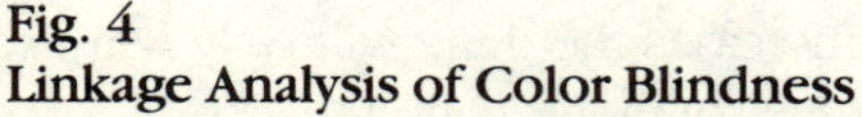

Fig. 4
Linkage Analysis of Color Blindness

Suppose that near the color-vision gene there is a region of DNA, which geneticists call a marker, that appears in two distinguishable forms called A1 and A2. The form, or allele, of the marker that each person has is shown on the family tree.

In this case, the connection with the marker is obvious: All the color-blind men have the A1 version, and all the ones who can see colors have the A2 version. This is because the color-vision area is near the piece of DNA we picked as a marker—so near that the pieces always travel together from parent to child. The marker is not causing color blindness, it's just being inherited along with the gene for color blindness.

In a different family, we might find that all the color-blind men have the A2 marker, but the ones who can see color have the A1 version. At first this seems like a flip-flop of what was seen in the first family tree, but it proves the same point: The identical form of the marker, no matter if it's version A1 or A2, always appears in people who are color blind. The consistent appearance of the same marker also proves there is a gene for color blindness. Remember, it's the gene that matters, not the marker. The marker can be anything and have a function totally different from the gene in question. A marker can be seen as a directional arrow: It doesn't matter what kind of arrow it is, as long as it's pointing in the right direction.

One way to understand linkage is to imagine that the entire genome is a city divided into twenty-three neighborhoods, representing the twenty-two pairs of regular, autosomal chromosomes plus the sex chromosomes, X and Y. The individual houses (genes) sit side by side on streets (strips of DNA). A marker is a specific block, so the marker called 700 block of Main Street would include all the houses (genes) on that block. The marker is not specific enough to pick out the individual houses, but it will always include the same houses. There are chemical procedures that allow scientists to make a probe for a particular block and pass it through the entire city, neighborhood by neighborhood, until it latches onto the block—the marker—it's looking for.

In LOD We Trust

In the family tree shown in fig. 4, the linkage between color blindness and the marker is obvious. But in science, obvious isn't good enough.

The significance of a linkage study is measured by its LOD score, which stands for logarithm of the odds ratio. The odds ratio is equal to the probability of the observed data assuming linkage, divided by the probability of the same data without linkage. To keep the numbers manageable, the sum is converted into the logarithm to the base 10. Thus a LOD score of 3 means there is only a one-in-one-thousand chance the linkage is caused by accident and is not significant. Most geneticists agree that a LOD score of 3 means a significant linkage between a marker and a trait, at least for Mendelian inheritance. For example, the LOD score for the color-blindness pedigree in fig. 4 is 3.4, which is comfortably above the usual cutoff. More encouraging yet would be a LOD score of 4, indicating odds ten thousand to one in favor of linkage, and so on up the scale. The higher the LOD score, the stronger the evidence for linkage.

Linkage for Complex Traits

Detecting linkage for complex traits is more difficult than for Mendelian characteristics because multiple genes and environmental factors can hide the connection. This was an important concept for us, because we suspected that the role of genes in sexuality could be subtle.

Even linkage for something that is obviously genetic, such as color blindness, can easily be obscured. For the sake of argument, imagine one of the uncles in our hypothetical family has another gene that cancels out his color-blindness gene. Then imagine that one of his nephews lied about being color blind to the researcher. Now the statistical analysis says there is no way that the color-vision gene is close to the marker, because the LOD score is less than zero and the evidence for linkage has disappeared. In fact, the linkage is still there, it's just hidden.

Sib-pairs and Shared Traits

Fortunately, there is a way to spot linkages that might be hidden. The technique is called the shared-trait sib-pair method. Unlike the classic studies of large families used to spot Mendelian inheritance, this method uses nuclear families and usually just two siblings, or a "sib-pair." Also, the search is narrowed to only those pairs who actually share the trait being sought.

For example, in families with two brothers who share the desired trait, in this case homosexuality, both brothers and the mother are tested for a marker that has two versions, called X1 and X2. If both brothers inherit X1 or both brothers inherit X2, they are scored as genetically concordant. If they inherit different marker alleles from their mother, they are called genetically discordant. The procedure is repeated on a large number of families, and the number of discordant and concordant pairs is calculated.

Our hypothesis is that the "gay gene" is linked to the X1/X2 region of the X chromosome. If our hypothesis is correct, then there will be more pairs of gay brothers who share the same marker and are concordant than pairs of brothers who have inherited different markers and are discordant. This is true regardless of whether the "gay gene" is located on the same chromosome as the X1 version or the X2 version of the marker. If our hypothesis is false, and there is no linkage, then the brothers will inherit the markers at random, and there will be equal numbers of concordant and discordant pairs.

The beauty of the shared-trait sib-pair method is its simplicity. According to Mendel's laws, each pair of brothers is like a coin: Each pair has an equal probability of being either concordant or discordant. Testing concordance is like tossing a coin many times over. If "heads" comes up about as often as "tails," the coin is fair. If heads comes up many more times than tails, the coin must be rigged in some way, in this case by genetic linkage. Just how rigged—how strong the linkage—is measured by the LOD score.

If the precise location of the trait gene is not known, it is simply a matter of trying different markers along the arms of the correct chromosome, in this case the X chromosome. If there is a "gay gene" somewhere on the X chromosome, eventually a marker will be found that shows linkage. If there is no such gene, then no such correlation will be observed. By testing enough markers—about 500 to 1,000—it's possible to search the entire genome for the genetic source of any trait that appears to have an inherited component.

In this method, only the DNA of definitely gay people is studied. An advantage of looking for linkage only in gay brothers and not in straight relatives is that "false negatives" are weeded out automatically. A false negative would be a person who has a homosexual orientation but identifies as heterosexual. "False positives" would be rare because heterosexuals are unlikely to lie and say they are gay.

Markers and Maps

Markers, then, can help find genes. But first the markers themselves have to be found.

Humans are nearly identical, genetically speaking. Only about 0.1 percent of the human genome, or three million DNA base pairs out of the total three billion, varies from one person to the next. (Base pairs are chemicals that bind together to form the rungs on the spiral staircase that DNA resembles.) Markers are found among the bits of DNA that differ from one person to the next. Such differences, called polymorphisms, can consist of deletions or insertions of genetic material or alterations in just a single base pair.

Identifying all these markers, these bits of DNA that vary from person to person, is the goal of the Human Genome Project. New markers are being added almost daily, making the genetic map ever more precise. Anyone with a computer and a modem can call Johns Hopkins Medical School and access the Human Genome Database, a sort of atlas of the genetic structure of the human body.

The map needs to be as complete as possible because of something called recombination. Every once in a while, two paired chromosomes will break and rejoin with one another. The result is that two strips of DNA that originally were on the same chromosome are now on different chromosomes and will show up in different germ cells.

For complex traits such as homosexuality, recombination is a major nuisance. The problem is that if a trait is only loosely related to genes, recombination will further weaken the relationship. For example, if homosexuality were linked to a very nearby marker on the X chromosome in two thirds of the people studied, it would take 30 pairs of brothers to get a significant LOD score of 3. But if the marker were located just 10 percent farther away, or about 10 million base pairs, it would require more than 80 pairs of brothers to prove the same linkage.

Chain Reaction

Stella Hu is a master with DNA, but her first attempts at linkage mapping were a disaster. When we started the experiment, most genetic mapping was done with a technique called Southern blotting, which even a few years later seems as quaint as making butter in a wooden churn. There were many opportunities for mistakes with this method, and at the rate we were going, we would have needed several years to test just a handful of genetic markers on our families.

One day I heard Stella having an animated conversation in the hall with a young postdoc from the lab of Wesley McBride, a senior scientist in the Laboratory of Biochemistry who has devoted his entire career to developing techniques for mapping human genes. I understand a fair amount of Chinese, but Stella and her friend were so enthusiastic about something that the conversation was moving too fast for me. All I could pick up were occasional bits of English words such as "trinucleotide repeat" and "PCR."

Afterwards, Stella explained to me that the postdoc was working on something called a simple sequence repeat polymorphism, or SSRP, which she called a "serp." These are short stretches of DNA in which two or three base pairs of DNA are repeated again and again. The useful thing about them is that different people have different numbers of repeats on their chromosomes. For example, on one chromosome the chemical sequence might be GGATTCACACACACACACACACACA-CATTAC, which contains eleven repeats of the sequence CA, but on another person's chromosome the same stretch of DNA might read GGATTCACACACACACACACACACACACACATTAC, which has thirteen repeats of the phrase CA. These sequences serve as genetic fingerprints found all over the genome, making it possible to map every bit of DNA.

From a practical point of view, the most important thing about these markers is they can be detected easily and economically with a method called the polymerase chain reaction, or PCR. Heating and cooling the DNA while mixing it with special enzymes and chemicals causes the short stretches of DNA markers to duplicate again and again. After forty cycles of heating and cooling, the original target DNA has a billion copies. The technique, once dismissed as an academic exercise with no practical use, has become one of the most powerful tools of molecular biology and won the inventor, Kary Mullis, a 1993 Nobel Prize.

PCR works because each of the four chemicals that form DNA can only make a rung on the ladder—a base pair—with one other chemical. So adenine (A) always pairs with thymine (T), and guanine (G) always pairs with cytosine (C). That means knowing the pattern of chemicals on one side of the ladder automatically gives you the pattern on the other side. Working like a chemical Velcro, only opposites will stick together to form the DNA ladder.

Stella's first try using the PCR method was a huge success. She set up the reactions in the morning, ran the gel in the afternoon, and had a readable X-ray film by 6 P.M. the same day. This was a big improvement over the old method and the start of a chain reaction of genetic mapping in our lab. Stella could run the reactions as fast as we could get her samples of DNA from our families, and that's where I came in.

Making DNA

I hate making DNA. Not only is it a long and tedious procedure, it also is potentially dangerous. Working with human blood always is dangerous, especially now with the AIDS virus. Our samples known to have HIV were sent to an outside lab for processing, but there might also have been samples contaminated with hepatitis B or some other unknown bugs. I always wore gloves, a lab gown, and safety glasses, and I had to break old habits, such as filling thin glass pipettes by sucking on one end. Even with all the precautions, it still was potentially harmful, and since the whole thing was my idea anyway, I figured I should be the one to make the DNA. In any case, I was the only one available, because Stella was getting ready to draw maps, Nan Hu was working on chromosomes, and everyone else in our little corner of Building 37 was working on their own projects.

I started out making DNA the old-fashioned way: brewing it up in a batch of toxic organic chemicals that had to stew for several days, just as I had learned twenty years before in graduate school. This produced a spidery mass of silky white threads of DNA, technically known as GSB, or giant snot ball. This was kind of fun, but it took forever.

One day, leafing through a technical journal, I came across a short article promising a more "rapid and economical" method of preparing the DNA. With the next blood samples, I decided to prepare half the old way and half the new way. As advertised, the quick technique took only a few hours and required few manipulations. But when I pulled the DNA out of the solution, it wasn't in long threads but in a clump, which of course was called TSB, where T = tiny.

I wondered if the two types of homemade DNA would work equally well for PCR mapping, but I didn't want Stella to laugh at my pitiful specimens. I decided to test the first batch myself, just to see how it worked. I picked a marker called DXS52 because its products can be seen without using radioactivity, which is the normal way of looking at DNA. I already was working with potentially lethal blood samples, so why add a daily dose of radiation to my job description?

I ran the PCR reactions and found good news and bad news. The good news was that the DNA samples made by the new method actually worked better than those prepared by the traditional technique, perhaps because they were less stringy. The bad news was that in all four families I was testing, both gay brothers seemed to be coming up the same, which meant the marker I had chosen might not be useful.

To be sure about the marker, I tried the experiment again, this time using the DNA from the gay brothers in the four other families I'd collected by this time. Once again, all four pairs seemed to be concordant for the marker. Fearing that the marker was too common to tell me anything, I decided to try one more test, using the marker on the DNA from the four mothers from whom I'd been able to collect blood. Somewhat to my surprise, I found that three of them had different forms of the marker on their two X chromosomes, which indicated the marker was indeed working.

The very next pair of gay brothers I tested came up discordant. They clearly had inherited different versions of the marker from their mother and therefore couldn't share the same "gay gene." I couldn't be sure whether there really was a link between the DXS52 marker and sexual orientation, and now I'd run out of families to test. I realized I was going to need more families and more markers. Finding more families meant going back on the road, but the markers I could find just by logging on to my computer.

I called up the genome database and found the marker DXS52 I had been using. On the map, it's located in a region called Xq28 at the very tip of the long arm of the X chromosome. The name of the region comes from the chromosome (X), the arm of the chromosome (q), and the position on the arm (28). When I looked to see what else I could use as a marker that might be located close by, I saw I was in a very well known neighborhood. Indeed, it had more known strips of DNA than any other place on the entire genome map.

Scrolling through the map of Xq28, I saw the gene that gives red-green eye pigment, or fails to and creates color blindness. There was the blood clotting factor VIII gene, which causes hemophilia if it is defective. Nearby was the glucose 6 phosphatase gene, which is deficient in a common metabolic disease in the Middle East. Many other traits, from diabetes insipidus to myopia to adrenoleukodystrophy (as in the movie *Lorenzo's Oil*), also had been mapped to this general region of the X chromosome.

It seemed unlikely any of the known traits linked to the Xq28 region would have anything to do with sexual orientation. The advantage of being in the area, however, was that it was so well known and that more and more markers were being discovered. There already were so many that in the course of several minutes and with just a few clicks of the mouse, I had the DNA sequences for three of the closest markers stored on a disk. These would help narrow our search for the gay gene, if there was one.

I put on my tie and sport coat again and left the lab to resume the search for additional families. I was confident that Stella and the others could test the new markers for Xq28, looking to see if there was any connection to sexual orientation. It was going to be a tedious process, because finding the right marker was a one-in-a-million chance, but we had to start somewhere, and since Xq28 was so well explored, I told Stella that it seemed as good a place as any.

It wasn't until a few months later that I realized how phenomenally, blessedly lucky I had been.

Chapter Seven

THE HARVARD CRUCIBLE

Professional interest in our research was mixed. A few colleagues were eager to see the results and thought there might be applications for their own work. Others were extremely uncomfortable with the topic and didn't want to come near my lab. So I was mildly flattered in March of 1992 to receive a letter from a Dr. Evan Balaban, an assistant professor in the Department of Organismic and Evolutionary Biology at Harvard University. Written on Harvard letterhead, with the little seal that proclaims "veritas," the letter explained Dr. Balaban was teaching a course on the genetics of behavior and that he'd like to use our project, which he'd heard about from a friend, as an example of "how modern behavior genetic studies are overcoming methodological pitfalls which plagued earlier research." He went on to say, "My interest is not in whatever results you may have obtained so far: I only want to share with the students the ways in which conceptual advances have been incorporated into modern family studies."

I was glad that Dr. Balaban wasn't interested in results, because we didn't have any. At that time, we had just begun the project, and I was spending my days on the road, interviewing subjects. I did send him a copy of a little pamphlet we were giving to volunteers, describing the project in very general terms, and a short note explaining that we were just getting started but that I would keep him apprised of the results. I had never met Balaban, but I was happy to help a professor who was trying to teach his students about the latest scientific work. I admit that I liked the idea of being the subject of a class at my alma mater. It felt like an imprimatur of some kind.

The next six months were busy with recruiting and interviewing gay brothers and their families, filling out paperwork, logging all the information, and listening, listening to stories about coming out, about sex—gay, straight, and in between—about AIDS and T cells and second cousins twice removed who might, just might, be gay. At first the stories were interesting, but after a few dozen versions of life's passages, all I wanted to know was "Are you gay and can I take your blood?"

I had left Stella in the lab, and she was making rapid progress. The freezer was beginning to fill up with boxes full of DNA, each sample in its own tiny tube identified only by a number. Stella had no way of knowing whose DNA she was testing, so there was no way for her to tell which traits were linking up to what markers. She didn't mind, however, and stood patiently day after day in front of a crowded workbench, doing the painstaking, tedious work of testing the samples.

The first step was to delicately place the DNA into small plastic trays shaped like muffin tins. Each tiny well held DNA from one person, plus a cocktail of primers, enzymes, and some radioactive material to make the DNA visible. Each tray then went into a special oven that pushed the temperature up and down about forty times to carry out the chemical reactions necessary to multiply the tiny segments of DNA.

Separately, Stella prepared a substance like gelatin and poured it ever so gently into the paper-thin space between two sheets of glass. When the liquid hardened into a gel divided by thin lanes, Stella carefully placed each DNA sample into a separate lane. Current was applied to the gel, causing the molecules to move toward the positive charge.

After several hours, she would delicately remove one sheet of glass and cover the exposed gel with a piece of X-ray film to pick up the images of the slightly radioactive molecules, now arrayed in order of size. The molecules, frozen as if they had been in a very slow race, left small dark bands on the film. When pairs of brothers shared the same pattern of bands, that meant they shared the same genetic marker. The mother's DNA also was tested to see which of her two X-chromosome markers had been passed to her sons. The marker was not the same as a gene, but it was a signpost to the gene. At this stage we were testing markers on individual families, not the whole group, to see if they had DNA regions in common. In other words, we tested the markers to see if the gay brothers in the "Smith" family shared a region, not if they shared a region with the gay brothers in the "Jones" family. That test would come later.

The question we were trying to answer was whether there was a connection between the marker and being gay. Finding such a linkage depended on three things. The first was whether there was a "gay gene" near the marker. If there was not, then no matter how many families we tested we'd never find a connection. A second factor was how many people were influenced by the hypothetical gene. If every gay man had the same version of the gene, we'd only have to test about 10 families to find a significant linkage and statistical significance or high enough LOD score. But if the gene played a role in a smaller fraction of gay men, we'd need to test many more families to determine the influence of the gene.

The third factor was how informative the markers were. If, for example, a gene had two "flavors," or variations, then half the time the mother would have the same flavor of both genes and it would be impossible to tell which of the two had been passed to the sons. In other words, if the mother had two genes called vanilla, the sons would both get vanilla, but there would be no way of telling which of the two vanillas. The ideal was a marker with as many flavors as Baskin-Robbins, each of which could be identified in the sons. In other words, the more variety the marker had, the easier it was to track its progression through the family tree. If the markers were coins, for example, using only pennies wouldn't tell much, because they look the same. If you use dimes, nickles, quarters, and pennies, it's easier to pick out the one you need.

We had started with just four markers, scattered over the X chromosome, and we kept testing additional ones until we had 19 markers, covering most of the chromosome. Each marker covered only a small bit of DNA, but combining them gave us broad coverage. By the time we had tested 21 families, I could make a ballpark estimate of whether any of the markers were linked to sexual orientation by counting how many pairs of gay brothers had the same alleles, or flavors, of the same genes and how many had different alleles. Because of the way the genetic dice are thrown, half the brothers will share a given marker just by chance. So if more than half share the markers, that means linkage is possible. Using the coin analogy, if the mother has an equal number of pennies and nickles to give to her sons, half the time they will get the same coin just by luck. If more than half of them get pennies, then it's possible the results are weighted. When I looked at the first 21 pairs, there was an obvious pattern. The markers at the end of the long arm of the chromosome, in the general vicinity of Xq28, were the only ones that continued to be shared more often than not.

Remarkably, one of the markers still showing linkage was DXS52, the very first one I had chosen—because it was convenient. When I had run the first tests myself, and the marker had hit on seven out of eight pairs of brothers, I had thought I was doing something wrong or that the marker didn't have enough variety. To have that many hits was like a coin coming up heads seven times out of eight: not impossible, but unlikely. Now we had 21 pairs, and most of them still were showing linkage for DXS52, plus linkage for two markers flanking it and another one a bit downstream. Somehow I had managed to throw a dart at a map of three billion base pairs and hit very near the bullseye.

There was no celebrating yet, however, because we didn't have a LOD score of 3, which was the statistical cutoff for good linkage. One problem was that for some pairs of brothers we were unable to get DNA samples from the mothers, either because they were no longer living or because they didn't know their sons were gay and weren't participating in the study. This meant that if the sons seemed to have the same marker allele, we couldn't be sure if they really had inherited the same region of the X chromosome or if their mother had only one flavor of the marker gene to pass on to her sons. If the mother only had pennies to give, seeing pennies in her sons wouldn't tell much. Instead, I had to make a probability calculation. At first, when our own collection of families was small, we used data from other labs that showed the frequency with which given alleles appeared in the population. As our own sample grew, we developed our own database on allele frequency, which prevented us from being thrown off the track by some unforeseen genetic idiosyncrasy of the families we were studying.

The next task was to figure out exactly where within the Xq28 region sexual orientation was mapping. To narrow down the search for the location of a gene as close to a marker as possible, we used a technique called multipoint mapping, which uses several markers to calculate the position of a gene. Fortunately, so many human geneticists were interested in Xq28 that the markers in this area already had been well characterized and pinpointed.

When I ran the numbers in a computer program called "Linkage," I got good news and bad news. The good news was the LOD score was more than 2, meaning we were approaching statistical significance. The bad news was the position of the "gay gene" was ambiguous. It could have been sitting downstream of our markers in a region called Xq27, or up at the tip of the chromosome in an area called the telomere. If we hypothesized that the gene was responsible for homosexuality in some families but not in others, that made it even harder to determine the exact location.

The only solution was testing more markers and recruiting more families. There were markers available in Xq27, downstream from Xq28, because this region contains the gene for a form of severe mental retardation called the fragile X syndrome, which was being mapped intensely in many other labs. Markers upstream were harder to find because there seemed to be little interest in this region.

We got a lucky break in November of 1992, when we were given access to two new markers in the most upstream part of Xq28. Ever since we had narrowed the search to Xq28, I'd been on the phone trying to find out who was working on this region and whether they'd found any new markers. One of my calls was to David Schlesinger, a well-known geneticist at Washington University in St. Louis. One of the designated mappers for the Human Genome Project who was working on Xq28, Schlesinger said a former student of his, Diha Freije, had moved down the hall to work with another renowned gene mapper, Helen Donis-Keller, precisely on developing markers for the portion of Xq28 closest to the telomere, exactly where we wanted one.

I called Donis-Keller and Freije, and they said they had indeed isolated two new markers in the upstream portion of Xq28 and had already used them to make a fascinating discovery. They found that at the very tip of the long arm of the X chromosome, which marks the end of the Xq28 region, there is a short stretch of DNA identical to the corresponding region of the Y chromosome. They had isolated one marker there and another just downstream of it.

The little region they had discovered normally stays with the rest of the X chromosome on the journey from father to daughter and with the Y chromosome from father to son. But every once in a while—in about 2 percent of male meioses (when the chromosome pairs split in the spermatocyte)—the X and Y chromosomes get jumbled up, and this little strip of DNA from a Y chromosome is "mistakenly" passed to a daughter (or a bit of the X goes to a son). That means boys are getting a tiny bit of a "female" chromosome and girls are getting a bit of a "male" chromosome. This raised the intriguing possibility that a genetic crossover between the male and female sex chromosomes is related to the behavioral "crossover" between heterosexuality and homosexuality. Even if that theory proved too convenient to be true, the markers were right where we needed them.

Freije and Donis-Keller had written up their results and submitted them to the journal *Science,* and they promised to send the details as soon as the paper was accepted for publication, so we could make the markers ourselves.

The manuscript came in the mail just in time for the arrival of the final member of our team. Victoria Magnuson had just received her Ph.D. in molecular genetics from the University of Texas Medical Center at San Antonio and was interested in learning the sort of detailed analytical techniques needed to pin down the connection between a genetic marker and sexual orientation. She was also interested in behavior genetics because her undergraduate training was in ethology, the study of the genetics and evolution of behavior in animals.

As soon as Vicki was settled in, I briefed her on the emerging connection between Xq28 and sexual orientation, and she eagerly sought still more markers to continue localizing the "gay gene" in the hope of ultimately isolating it. When she read the Freije and Donis-Keller manuscript, she too recognized its theoretical and practical significance. Within a few weeks, she'd set up her own mapping operation and was busy going through our families with the new markers.

As the new markers were tested on each person, and as more and more families were added, the amount of data began to grow exponentially. Soon I couldn't even run the analyses on my lab computer: The memory wasn't big enough. Fortunately, Vicki had substantial training in computers and quickly figured out how to run our numbers through a larger machine at the NIH. Even this wasn't enough capacity, however, as I learned from an irate telephone call informing me that we were tying up more memory than was allocated for our entire division of the Cancer Institute. To keep us out of trouble, Vicki tapped into the Frederick Cancer Center's Biological Supercomputer Facility, which has a mainframe computer big enough to handle the billions of calculations we needed.

By March of 1993, we had typed 36 families with 21 different markers that spanned the X chromosome from one end to the other. A quick run of the data showed the only obvious linkage still was showing up in the vicinity of Xq28, but we planned to test more people and more markers.

I was so engrossed in the experiments, I had forgotten all about the letter from the Harvard professor, Dr. Balaban, saying he wanted to teach a class about our project. But I had received a letter, dated 16 October 1992, from Ruth Hubbard, a professor emerita from the Harvard Biological Laboratories and one of the best-known—and feared—people in the business. When I was still a graduate student at Harvard Medical School, she had led the fight against recombinant DNA, a technology she claimed would allow "cancer-causing bacteria" to run amok through the streets of Cambridge. The city for a time even banned such research, which meant the local labs were left behind other labs around the world. Twenty years later recombinant DNA–produced drugs, such as a safe vaccine for hepatitis B, are saving human lives worldwide, and the economy of Cambridge is heavily dependent on the recombinant DNA–based biotechnology industry.

Dr. Hubbard stated that she had read the little pamphlet I had sent to Harvard and found our proposal "scientifically unacceptable." This seemed like judging a marriage by reading the wedding announcement. In a caustic aside she added, "Indeed, I understand that one of my colleagues at Harvard is using it in a course as an example of how not to do linkage studies."

That was too much. Criticism from Hubbard was to be expected, but it was more upsetting to have been blindsided by some Harvard professor I had never even met, a professor who apparently passed judgment on our project without even knowing what we were doing or what the results were. I imagined him waving our pamphlet and teaching his students—students who were supposed to be learning to make judgments based on sufficient data—that we were doing bad science. I threw down the letter and picked up the telephone to call Hubbard. The conversation started out pleasantly enough, but I was upset, and she isn't known for being shy about expressing her opinions. She refused to tell me the name of her "colleague" who was teaching about our project but agreed to have him write me a letter. A few days later, Hubbard wrote back to say she was sorry she had angered me but insisting that, "Across the board, I find biodeterminist explanations unsatisfactory on scientific as well as political grounds."

I wasn't sure I ever would hear from her colleague, but I figured it had to be Balaban, the one who had requested the information on the research. I was surprised a few weeks later when I received a letter from none other than Richard Lewontin, the Harvard professor who had, quite unwittingly through his political manifesto *Not in Our Genes,* helped inspire me to launch the project in the first place. Lewontin was teaching the behavior genetics course with Balaban, and he said that they had indeed used our little pamphlet (which, as you will recall, they had solicited as an example of "conceptual advances" in human behavior genetics) as the basis for criticizing the entire field.

Since no one had been able to replicate some of the early psychiatric genetic linkage studies, Lewontin wrote, linkage studies in general must be unreliable. He went on to theorize that human behaviors must be "very, very far from the genes" because "there are some at least that we know for sure are not influenced by genes as, for example, the particular language one speaks." That made about as much sense as saying that since some people eat tacos and some eat hamburgers, there is no biological drive to eat.

When my blood pressure had more or less returned to normal, I realized the trap I had fallen into might actually work to my advantage. I immediately wrote back to Lewontin, saying, "I suppose I should be flattered that you are teaching about a study of mine in a course at Harvard University," but that "I was a trifle surprised to learn that you are already criticizing the study before it has been published and without knowing the exact design of the study, the methodologies employed, nor the results. If and when our work is published, I hope that you will be more open-minded."

In a final dig, I suggested sarcastically that if he should ever decide to teach his students how a behavioral genetics study *should* be done, I'd be more than happy to deliver a guest lecture.

To my surprise, Lewontin took me up on my offer. He wrote back and said, "It might be extremely interesting, in fact, to confront the students with someone who is actually doing this sort of work." He made it sound like I was some sort of evil mad scientist, and I imagined him eagerly readying the pillory to humiliate another heretic, but it was a challenge I couldn't pass up.

When the day arrived, I stood up in a classroom of the old Agassiz Museum of Comparative Zoology, which looks like a cavernous attic filled with dusty skeletons and animal exotica floating in formaldehyde. Some people might consider me brave to appear in the inner sanctum of my inquisitors, but I'm not stupid. I had invited a few witnesses for the defense, including Richard Pillard, who teaches at nearby Boston University and who was one of the people who helped me prepare the original proposal; and Chandler Burr, a journalist who had just published a widely discussed *Atlantic* magazine cover story called "Homosexuality and Biology."

The other side was led by the curly-haired Balaban. The tall and tweedy Lewontin was there, too, as well as a few of his disciples. Seated in the front row was Ruth Hubbard herself. Also in the audience were forty students, the supposed reason for my lecture. In fact, the students were mere spectators for the proceedings, although it was a little disconcerting to think they had been taught by their professors that I was, basically, an idiot.

I focused my talk on the criteria for *any* good human behavioral genetics study rather than on the details of our own work. I stressed the importance of clearly defining phenotypes, the need to look directly at the genetic information itself instead of the old method of relying only on family studies, and the requirement for high-resolution genetic maps to detect genes that influence, rather than determine, complex behaviors. I had a stack of references culled from the theoretical literature showing how the role of genes could be teased out in complex behaviors by looking at siblings who share the trait in question. Only at the end of the talk did I describe our own data indicating a linkage between male sexual orientation and Xq28.

When I finished, I asked for questions, bracing myself for what I expected would be a hostile barrage of criticism, perhaps a few hoots of derision, and a generally unpleasant thirty minutes. Instead, Lewontin stood up and graciously announced to the class that I had addressed essentially all of the concerns he and Balaban had had about human behavioral genetics studies. As far as he was concerned, Lewontin said, our study was scientifically sound. I had been forgiven. I wasn't sure if I should kiss Lewontin's ring, but he ducked out of the lecture hall without speaking directly to me.

Dr. Hubbard stayed to ask one rather bizarre question about hermaphrodites, which I think I answered. I knew she wasn't through with me yet, though, but was saving her criticism for another day. She was working on a book of her own called *Exploding the Gene Myth,* and my research would take a place of dishonor in her exposé of "how genetic information is produced and manipulated by scientists, physicians, employers, insurance companies, educators, and law enforcers."

I had taken a risk during the lecture by confidently, and for the first time publicly, talking about a link between sexual orientation and

Xq28. At that point, the final numbers still were being worked by the computer in Frederick, so when I returned to my hotel room, I connected my laptop to the telephone and dialed up the latest results.

The analysis was complete, and the results were clear: Sexual orientation continued to map at Xq28, and the multipoint LOD score had exceeded 3.0, the usual cutoff considered to be statistically significant. This new data meant we could exclude the downstream Xq27 region, and it meant there was only one chance in one thousand that the linkage we had found was an accident.

Now we were getting somewhere. Looking at the patterns of inheritance in our families had suggested there was a genetic link to homosexuality and that it was located on the X chromosome, but that was an inference rather than a proof. Now by studying DNA we'd found a connection between homosexuality and the chromosome itself. We didn't know what the gene did or how it worked—nor, for that matter, exactly which gene it was—but we had strong evidence that it existed. This was terrific news to take back to the lab, and I wondered if it was time to start writing up the results for publication.

Chapter Eight

GOING PUBLIC

Any hesitation I had about publishing our results was blown away by a telephone call from a reporter. She must have had very good sources, because she knew all about my talk at Harvard and that we had found a genetic link to homosexuality. She wanted to interview me for a story, but I was reluctant. In many ways it was too early to go public with our work: We hadn't actually isolated the "gay gene," only detected its presence, and the results would need to be replicated and confirmed on an independent sample. On the other hand, we did have support from both family and DNA studies, and a statistical significance in our study that would be recognized as solid by people in the field. The reporter's call raised the possibility of the worst of both worlds, which would have been someone else making our work public, possibly out of context and without all the supporting evidence.

The reporter didn't care whether we were ready or not. She smelled a story and wanted to run with it. I managed to put her off for a time, but I knew I had to act. The alternative was to be swept away in media hoopla before the work had gone through a rigorous peer review and been published in the scientific literature. Too many scientific "discoveries," such as cold fusion, had been announced with great fanfare by the media only to fall apart under even modest scrutiny. I didn't want that to happen to us.

I decided to try to publish in *Science,* the weekly publication of the American Association for Advancement in Science. It is the most widely read scientific journal in the world and covers all areas of the natural sciences, from biology to astrophysics to geology. The scientific standards are exceptionally rigorous, and every paper must go through two layers of review, first by an editorial board to determine general suitability and then by a panel of experts to judge the technical merits. The competition for space is so intense that a negative word from just one reviewer is enough to cause a paper to be rejected. The real advantage of publishing in *Science* was that the paper would receive the most rigorous possible scientific and technical review, and publication would spur other labs to replicate the work—either supporting it or debunking it.

THE PAPER

When I wrote papers on my old topic, metallothionein, the standard opening line was something like "A large body of compelling evidence demonstrates that . . ." When I sat down to write the paper on sexual orientation, I realized there was precious little evidence at all, much less any that the readers of *Science* would find "compelling." I opted instead for a straightforward description of what we did know: "Human sexual orientation is variable." That seemed hard to dispute. Then, since the paper was about the role of genes and biology in sexuality, I gave a brief review of previous family, twin, and brain studies, leaving it to the reader to decide the significance of the evidence.

Most papers on modern human genetics are about diseases or traits that have been widely studied, so the authors don't have to spend a lot of time on what the trait is and how it is identified. Rarely does a single paper try to cover a description of the characteristic, the family pedigree studies, and the linkage data. But we were in virgin territory, so to speak, and had no choice but to address all three areas at once. I wanted to avoid putting too much emphasis on our family data, because I don't think family trees alone can prove a role for genes, but I couldn't ignore the family data, because that was what pointed us to the X chromosome.

When I got into the linkage analysis, I was at a loss for words. All the linkage papers I'd ever read were about diseases and talked about "disease genes" and "patients" and "illness," terms that were not appropriate in a discussion of sexuality. Then I wasn't sure about the words "gay" versus "homosexual," or even whether to use them as nouns or adjectives. In the end, if our paper accomplished nothing else, it introduced the word "gay" into the official lexicon of the American Association for the Advancement of Science.

Finally, there was the ending. I felt strongly that if I was going to get the credit (or blame) for discovering a genetic link for sexual orientation, I had a personal duty to do everything possible to see the information was used in an ethical and responsible way. What better place to start the necessary discourse than in the same paper announcing the scientific results? All too often, ethical considerations have lagged behind rather than developing hand-in-hand with the science. So I ended our paper with the caution that using genetic research on sexual orientation or other human attributes to discriminate against people would be unethical. I knew that not everyone would agree, but I felt an obligation to raise the issue.

After going over the first draft with my lab colleagues, I did what I normally did with any paper: I sent it outside the lab for initial reactions and feedback. When I wrote about metallothionein, I'd give the paper to two or three experts in the field—any more and there might be no one left to review it for the journal. I knew this paper was going to be more closely scrutinized than anything I'd ever written, so I decided to have it reviewed informally by a full baker's dozen of experts: six geneticists, three statisticians, two molecular biologists, and, against my better judgment but because I knew I had to, two psychiatrists.

The manuscripts came back with a number of comments on style (two reviewers didn't like the word "gay"), statistical analysis (several wanted a higher LOD score—as did we), and content (most thought the ethical comment was "inappropriate"). No one, however, questioned the basic scientific approach or the way we interpreted the available data. I substantially revised the manuscript with the suggestions in mind, although I kept the ethics paragraph. I sent the final version to the editors of *Science,* who would assign experts outside the journal to determine if the article was worthy of publication.

Six weeks later, the three reviews from *Science*'s outside experts came in over the fax machine. I was almost afraid to look. I knew that the identities of the reviewers were kept secret, so they could be blunt without me blaming them personally. I stood there with my mouth open while I read the opening of the first review: "Hamer et al. present a tour de force of modern human genetics, one that is likely to be considered by future generations as one of the great landmarks in the study of human behavior . . . By using genetic methods, this study is far more significant than the earlier neuroanatomic studies because the anatomic approach does not resolve the question of cause and effect —a genetic study by its very nature gets at the initiating event in the causal chain."

Even with that glowing opener, the reviewers had several suggestions. The first wanted us to be sure there wasn't anything unusual about how DNA is inherited at Xq28 that could have biased our results. Fortunately, I'd already ruled that out with another test. The second reviewer wanted us to emphasize that the results showed that the linkage was related to sexuality but not necessarily a cause of sexuality, which was a point of view I shared. The third made a strong argument that we should reanalyze the linkage data by a different statistical test. This took another week of computer programming and calculations, but in the end the numbers all agreed out to the sixth decimal place.

None of the reviewers required any more experiments, but we had continued to interview families and examine their DNA. By the time the paper was resubmitted with the reviewers' suggestions incorporated, the LOD score had edged up to 4.0, which meant the odds were ten thousand to one that the linkage was real. We also had completed enough markers to give us a finely detailed map of not just the Xq28 region but of the entire X chromosome. After one final round of review and revision at the journal, I heard from a senior editor on 17 June that we'd been accepted. The editors realized it would be difficult to keep the story under wraps for long after circulating it to so many reviewers, so they put it on the "fast track" for publication in one month. On 16 July, Volume 261 of *Science* carried our six-page research article entitled "A Linkage Between DNA Markers on the X Chromosome and Male Sexual Orientation," along with a separate article by a journal writer about the research. The original article appears in full in Appendix A.

THE TAKE-HOME MESSAGE

The research article outlined how we started with a simple question: Is there a gene on the X chromosome that influences male sexual orientation? Then we devised a straightforward test: If there is such a gene, there should be markers close to it that are shared by more than half the pairs of gay brothers in our study. Half the brothers would get the marker by chance, so we would have to find it in more than half for there to be some genetic linkage. Finally, we got a clear answer: There are such markers, so there must be such a gene.

The complete experiment tested 40 pairs of gay brothers for 22 different markers along the X chromosome. The major finding was that 33 out of these 40 pairs were concordant, or the same, for a series of 5 markers in chromosome region Xq28. Thirty-three out of forty is 83 percent, which is considerably more than the 50 percent chance that a marker would be shared without having a connection to the brothers' homosexuality. Mathematically correcting for the fact that half the pairs would be expected to share even random markers, we calculated that 67 percent of the brothers were "linked to" the Xq28 region.

The statistical analysis showed that the chance the correlation between sexual orientation and the markers at Xq28 could have arisen just by the luck of the draw was one out of one hundred thousand. The multipoint mapping gave a LOD score of 4.0, meaning it was ten thousand times more likely there was linkage at Xq28 than no linkage. To further test the statistical significance, we calculated the likelihood of our data assuming we had tested *every* possible marker in the human genome. Even by this rigorous standard, the chance we were wrong was only 0.5 percent. No matter how the data were analyzed, there was a better than 99 percent probability that the observed linkage was real.

Our second finding was that sexual orientation was *not* strongly linked to any other region of the X chromosome. Specifically, for all of the 16 markers tested outside of Xq28, the LOD score was less than 0, meaning it was more likely the markers were not linked than that they were linked. Although this was a negative result, it was important for two reasons. First, it served as an internal control, showing we were doing the genetic mapping the "right way." Second, finding that there was no linkage to other parts of the X chromosome meant that researchers who wanted to replicate or expand on our work wouldn't have to bother with those regions of the chromosome.

The third major finding was that gay men have more maternal than paternal male relatives who also are gay. In retrospect, the family data not only helped tell us where to look, it also supported the linkage results. If we'd found an X-linked marker but had not observed more maternal than paternal gay relatives, the DNA results would not have made sense.

QUESTIONS, CAVEATS, AND CRITICISMS

As I had expected, the *Science* article stirred up plenty of controversy. The scientific reaction included some criticisms that can be addressed easily and others that will take years to answer. Six of the specific scientific questions will be discussed here, plus four things the study failed to show. The more explosive popular reaction to the study will be discussed in the final chapter.

Segregation Distortion

Our entire analysis was based on Mendel's first law, which predicts that each brother has the same fifty-fifty chance of receiving either copy of an X-linked marker from his mother. But what if Mendel's law of independent assortment didn't hold for Xq28? There is a rare phenomenon called segregation distortion, in which one copy of a particular chromosomal region is more likely than its partner to be passed on to offspring. If this happened at Xq28, our results would be badly distorted.

Other geneticists have long been studying Xq28-linked traits, such as color blindness and hemophilia, without ever seeing any hint of segregation distortion. To address this issue more systematically, we used an international database that contains linkage information on a series of large, three-generation families. Running calculations on these families, we found nothing unusual about the genetic behavior of Xq28 that would have biased our results. As shown in fig. 5, the randomly selected brothers from the database shared markers almost exactly 50 percent of the time. Only the gay brothers showed the lopsided inheritance indicating genetic linkage.

Missing Mothers

We had DNA available from only 15 of the mothers (and from one sister) of our gay brother pairs. How could we know whether the "motherless" matching brothers really had inherited the same Xq28

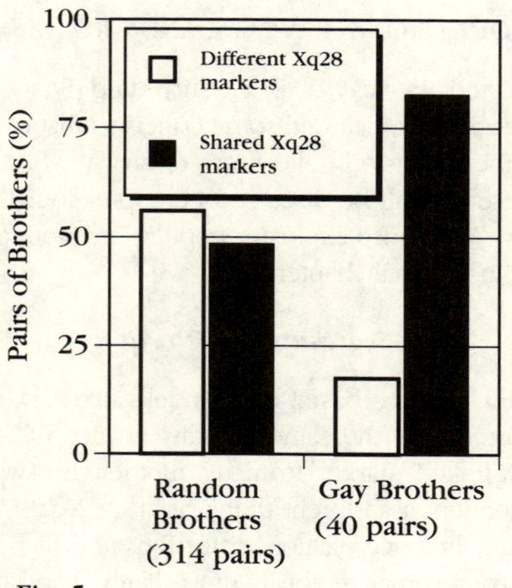

Fig. 5
Summary of Linkage Data

region, or if the mother only had one flavor of the marker to pass on to her sons? Fortunately, Xq28 is so well mapped we didn't have to rely on a single marker. We had five different markers that were so close to one another that they all could be treated as a single extended marker with 2,880 variations. This made it unlikely to get an erroneous result because the probability of a mother having only one version of all five markers was less than 0.5 percent.

In experiments after the paper was published, Vicki Magnuson typed our families for an additional four markers. The new data didn't change the results for a single pair of brothers, meaning the additional work strengthened rather than weakened our basic conclusion. In principle, it would have been better to have obtained DNA from every mother, but this was not possible because often the mother had died or did not know her sons were gay.

Heterosexual Brothers

A more frequently heard criticism was that we did not test the markers on the straight brothers in our families. In a *New York Times* op-ed piece titled "False Genetic Markers," the unbendable Ruth Hubbard wrote: ". . . the researchers did not do the obvious control experiment of checking for the presence of these markers among heterosexual brothers of the gay men they studied. It is surprising that the correlation found in this research warranted publication without these controls, especially in as influential a journal as *Science.*" Dr. Hubbard's criticism was surprising, particularly in light of the materials in which we provided her. The references I gave her showed why we deliberately used only gay brothers. The whole point was to test only gays in order to detect genes that swayed sexual orientation even if their influence was subtle.

There were three good reasons why we focused on gay brothers. First, people who say they're gay are unlikely to be mistaken, whereas those who claim to be straight are more questionable. Second, incorporating both heterosexual and homosexual brothers in our analysis would have required simultaneously estimating both the percentage of gay men who are gay because of the gene and the percentage of all men who have the gene and are gay. Every scientist knows it is impossible to find a unique solution for a single equation with two unknowns. The third reason was that including heterosexual brothers made sense if, and only if, we thought the Xq28 locus alone makes people gay. Ironically, Dr. Hubbard herself stressed the complexity and variety of human sexuality, but then expected us to run a test that assumed being gay was determined by some kind of tiny on-off switch made of DNA.

The real danger of doing the analysis Dr. Hubbard called "obvious" was that it might have missed a linkage that really was there. In a previous chapter, the example of color blindness was used to show how researchers could be thrown off the track of a gene if people who could see colors were lumped into the analysis. Then the linkage could be camouflaged by inaccurate assessment, incomplete penetrance, or interactions with other genes or the environment.

Obviously, heterosexual men should be tested, but not in the way Dr. Hubbard suggested. Checking pairs of brothers where one or both were straight would not be useful because of the low content of genetic information in such families. A better approach would be to look at heterosexual men with two gay brothers who were concordant for Xq28, which means the family was "loaded" for the gene. In early 1994, we were testing such families, and the preliminary results were encouraging. Of the first three heterosexual brothers tested, all had the opposite version of Xq28 from their gay brothers. The sample was too small to be significant, but it did seem to be pointing in the right direction.

Statistics

Genetics is a statistical science. There are people, however, who just don't trust statistics. One of them is Evan Balaban who, in an appearance on the "MacNeil/Lehrer News Hour," criticized the statistical analysis of our data on the grounds that ". . . it is probably possible, from many years of data, to come up with some significant correlation between stock market prices and phases of the moon." Ruth Hubbard has a similar problem. The example that she used in a letter to me was genetic correlations "of family members who vote Democratic and those who vote Republican."

I confess that we did not check for correlations between homosexuality and phases of the moon. Hubbard's sarcastic comment, however, inspired us to run a useful control experiment. Because much of the study was conducted during the 1992 presidential campaign, the subject of the election came up in most of the interviews, and subjects often mentioned their party affiliation in passing. This allowed us to test the X chromosome for party affiliation rather than sexual orientation.

Even by testing 22 markers, we did not find, as Dr. Hubbard had suggested, any "linkage" to political parties; all the LOD scores were well below zero. Obviously, if statistical techniques are applied badly, they can lead to spurious findings, but that is different from saying that all statistics are spurious.

The "Over-loving Mother Gene"

John Maddox, the editor of the prestigious journal *Nature,* the British counterpart to *Science,* wrote a lead editorial with the provocative title, "Wilful Public Misunderstanding of Genetics." He argued:

> What if it is accepted, for example, that it is true what the psychoanalysts say, that male homosexuality is in part determined by the influence of an over-loving mother? And what if the gene located at the end of the X-chromosome does not determine male homosexuality, but instead plays a part in telling whether a mother is "over-loving" in the appropriate sense? Then the gene concerned would be strictly irrelevant to the causation of male homosexuality, whose determinants would remain those of nurture rather than nature.

I was disappointed to see such a basic error in such a good journal, especially in an article about misunderstanding genetics. The problem with the "over-loving mother gene" is that it violates Mendel's first law of independent assortment. Suppose there really were a gene that made mothers "over-loving," or in any way more likely than average to raise gay children. In that case, there might be a linkage between sisters who had gay offspring, which is something we didn't look for in our study, but there would not be any linkage between the gay brothers, which is what we found. Since the gene would only work in the mother, there would be no reason for it to be shared by her sons.

Replication

As with any first-time study, our results need to be repeated, by us or another lab, before they can be fully accepted. Had we announced linkage for some common disease, such as manic depressive illness or Alzheimer's, labs around the world would have been repeating our experiments on their own samples the minute our paper was published. But the genetic study of sexual orientation was brand-new, and nobody else in the world had collected DNA from a comparable series of gay brothers and their families. Just as I had hoped, the publication of our paper stimulated new interest in the genetics of sexual orientation. In 1994, at least one group was collecting families, and two more were applying for funding.

In the meantime, our lab continued to seek families to repeat the entire linkage analysis. People sometimes ask if it isn't boring to repeat the same experiment. I just remind them of how Nobel laureate Alfred Hershey, one of the discoverers of the biological role of DNA, defined heaven: "To find one really good experiment and keep doing it over and over."

WHAT THE STUDY DID NOT SHOW

Proving that a gene exists is one thing. Isolating or finding the gene, measuring its incidence and effect on people, and understanding how it works is another, and those goals may take many years to achieve. Given the brief history of the genetic study of sexual orientation, we're pleased to have made as much progress as we have, but many things, indeed most things, remain to be explored.

Population Parameters

Our study was not designed to address the role of the Xq28 locus in the population at large or even among all gay men. Mary-Claire King, discoverer of the breast-cancer locus, made this point most succinctly: "It is impossible to use either the family history or the linkage data to estimate the magnitude of genetic influence on homosexuality in the general population."

King then outlined some of the questions that remain to be answered. What fraction of all gay men carry an Xq28-linked allele that influenced their sexual orientation? What is the frequency of this allele among heterosexual men? How many different alleles are present at Xq28, and what is the effect of each? What other genes and nongenetic factors influence sexual orientation, and what role does each play?

Our study could not address these questions because we deliberately studied a population enriched for the gene we were searching for—enriched because we only studied those families with two gay brothers, no gay fathers, and no more than one lesbian. This selected population was necessary to determine whether a gene on the X chromosome had any effect on sexual orientation, but it could not show the precise magnitude of that effect.

To estimate the role of Xq28 in the population at large, researchers would have to isolate the gene, determine what parts of the DNA sequence were variable, and then perform a population-based survey on a large number of individuals with various sexual identifications. The results of such an experiment on a particular population, say residents of Washington, D.C., still would be limited and would not necessarily apply to other populations, such as Salt Lake City or Tokyo, where the same genes might express themselves differently because of the different cultural environments. A person with a "gay gene" growing up in a repressive society, for example, might develop differently than someone with the same gene growing up in a permissive society.

This type of measurement is still a long way from being feasible. In the meantime, we can make only educated guesses about the importance of Xq28 in the population at large. On the high side, the region couldn't possibly influence more than 67 percent of gay men, the proportion "linked" to this region in our highly selected group of gay siblings. On the low side, if much of homosexuality is caused by environmental factors, or by a large number of interacting genes, Xq28 could account for as little as a few percent of the variation in male sexual orientation. The median range, taken from our linkage data and from the available twin and family studies, suggests that Xq28 plays some role in about 5 to 30 percent of gay men. The broad range of these estimates is proof that much more work remains to be done.

In-betweens

We never did find many bisexual men, so there is no way to speculate on the role of Xq28 in people who identify as bisexual, asexual, or anything outside our simplistic dichotomy of "gay or straight." This was not an oversight but deliberate, because our first goal was to determine whether genes had any influence on sexual orientation, which meant it was important to study only those individuals whose orientation was unambiguous. A more systematic approach to this question will be possible after the gene is identified.

Lesbians

Twin and family studies suggest that female sexual orientation is as likely to be inherited as male sexual orientation, but narrowing the role of individual genes will require more effort than did the male project. In general, women display a wider diversity of sexual expressions and developmental patterns than do men. Furthermore, by early 1994 we had not observed any obvious transmission patterns in our ongoing studies of the extended families of lesbians. A complicating factor was that even if the X chromosome were involved in female sexual orientation, the expected pattern would not necessarily show an excess of either paternal or maternal gay relatives, because a woman inherits one X chromosome from her mother and one from her father.

There is no way to be sure yet, but it's unlikely the same version of Xq28 associated with male homosexuality also is associated with lesbianism. If it were, we would have found more families with both gay men and women. If Xq28 influences sexual orientation by directing a person's sexual attraction, it would be unlikely to influence both gay men and lesbians, because the objects of their affection are just the opposite. On the other hand, if Xq28 acts more indirectly, perhaps influencing personality traits, it's conceivable the same region is involved in both men and women. During early 1994, Angela Pattatucci continued collecting families with lesbian sisters and obtaining DNA samples to begin to answer those questions.

A Linkage Is Not a Gene

The most important limitation of our research was that we didn't isolate a "gay gene"; we only detected its presence through linkage. We narrowed the search to the neighborhood, the X chromosome—and even the block, Xq28—but we didn't find the house.

More precisely, our mapping showed the "gay gene" is most likely somewhere between the markers GABRA3 and DXYS154, which span a distance of about five million base pairs. That represents less than 0.2 percent of the three billion base pairs in the human genome, but because the genome includes somewhere around 100,000 different genes, such a small area has enough room for up to 200 or so different genes.

Narrowing a search to two hundred apartments in a big city isn't very helpful if you are looking for one person. The traditional method for localizing a gene is recombinational mapping, which takes advantage of the rule that the number of crossovers between a gene and a marker is proportional to the distance between them. Unfortunately, this method is poorly suited for mapping complex traits such as sexual orientation. It only works well for simple, single-gene traits that follow Mendel's laws. The main problem is not knowing whether a gay man is missing a particular marker because of a recombination event (the mixing of DNA) or if he is gay because of some different gene or a nongenetic reason.

A second method of gene mapping uses something called linkage disequilibrium, which involves searching for patterns of markers among unrelated gay men instead of searching for sharing of markers among brothers. Such patterns would emerge if the "gay gene" were sitting very close to a distinctive set of richly varied markers.

The third approach is candidate gene analysis, which examines the coding sequences of genes that "look like" they might have something to do with sexuality. For example, genes that are expressed in regions of the brain thought to control sexual activity, such as the hypothalamus, or that act differently in men and women. Unfortunately, we don't yet know of any such genes in Xq28.

The final approach is brute force; that is, to look for sequence variations in each and every protein coding sequence in Xq28. Right now, we don't have the necessary knowledge, techniques, or people to do that, but thanks to the Human Genome Project, there should be a complete catalog of all the genes in Xq28 available sometime in the next five to fifteen years. DNA sequencing technology is developing so rapidly that we probably will have the necessary methods to screen all these genes for interesting variations. So if there really is a "gay gene" in Xq28, it will be found.

Chapter Nine

BIOLOGICAL MECHANISMS: GENES, HORMONES, AND THE BRAIN

The first part of this book was grounded solidly on facts: Most men identify themselves as either gay or straight. Each one either has a gay brother or he doesn't. The brothers share the same genetic markers at particular spots on their DNA or they don't. Those are the facts that led to the conclusion that there is a gene at the tip of the X chromosome that influences sexual orientation, at least in some gay men.

This part of the book must step away, for a moment, from that comfortable solid ground in order to address some of the more interesting questions raised by these facts. For example, what could a gay gene on the X chromosome do? How did it get there? Why hasn't a gene that presumably works against reproduction been eliminated by evolution? These questions cannot yet be resolved by facts. Only theories can suggest possible answers and ways to find them.

Theories can be valuable. Sometimes a theory will accelerate research by suggesting the next logical experiment. For example, if we suspected that the Xq28 locus influenced the metabolism of hormones, we'd start by looking at genes that regulate hormone synthesis or degradation. If we thought that a gene in Xq28 acted on a specific part of the brain, such as the hypothalamus, we might first look at genes that are active there. Theorizing also works as a reality check. If an observation is so bizarre that it seems to defy explanation, perhaps the observation should be rechecked. That's what happened with the much ballyhooed announcement of cold fusion; it really stumped the physicists, until they realized that the observation itself was wrong.

Theories can be dangerous, however. They can lead into blind alleys or dead ends—something that has happened often in research on sexual orientation—especially if they aren't checked against all the available facts. So theories can be insightful or misleading, reasonable or silly. The main thing to remember is that they're just theories. The good theories, as the philosopher Karl Popper pointed out, have two characteristics: They account for the maximum number of observations with the minimum number of assumptions, and they make predictions that can be tested.

MECHANISMS: AN OVERVIEW

How might a gene at Xq28 influence the direction of sexual attraction and behavior? This question really has two parts. First, what instruction is encoded, or produced, by the hypothetical gene? For example, the gene might direct the synthesis of an enzyme that converts one hormone into another. This question can be answered only by isolating the gene and decoding its DNA sequence, something that hasn't been done yet.

The second part of the question asks: Which pathway is the gene involved in? It might directly influence a person's sexual fantasy, or it might make a person more likely to acknowledge "unusual" fantasies. This level of the question is a little easier to address. One way we tried to answer it experimentally was to compare the sexual and psychological profiles of the gay brothers who shared, or didn't share, Xq28 DNA markers.

The idea that gay men are "like women," and lesbians are "like men," is one of the oldest, and seemingly most obvious, concepts in sexology. The only problem is, it's not true. In most respects, gay men still are "real men" and lesbians still are "real women." The notion has a certain appeal, however, because like most women, gay men are attracted to men, and like most men, lesbians are attracted to women. For that reason it's not surprising that much of the biological research on male homosexuality has looked for differences between men and women and then tried to see if those differences exist in some form between gay men and straight men.

In trying to understand the biological mechanisms by which Xq28 might influence the differences between the sexes, we'll examine four theories. There is no guarantee that any of these possible explanations for how a "gay gene" might work is accurate—and the correct answer might be something we never dreamed of—but they do open up ways of thinking about mechanisms.

The "Master Gene"

The genetic makeup of men and women is remarkably similar, and of the 100,000 or so genes that make us human, only a single one is responsible for virtually all the biological differences between the sexes. This gene, called the testis determining factor, or TDF, is located on the Y chromosome.

Since the Y chromosome is the only one of the forty-six chromosomes that is found only in men and not in women, it was an obvious place to look for differences between the sexes. Scientists found that where, in rare cases, a person had just one X chromosome or three X chromosomes, the person developed as a female; by contrast, people with two X chromosomes and a Y, or two Ys and two Xs, developed as males. This indicated that it was the presence or absence of the Y chromosome, rather than the number of X chromosomes, that differentiated men from women.

Still unexplained was why every once in a great while a person with two X chromosomes was male or a person with a Y chromosome was female. Closer inspection revealed that the XX males actually had a tiny bit of the Y chromosome that had transferred to one of their X chromosomes, and that the XY females lacked a small but essential portion of the Y chromosome. After patiently tracking down some of these individuals, geneticists found that just one small region of the Y chromosome was responsible for sex determination. By using the tools of molecular biology to carve out the precise DNA sequence of this region, they found it harbored a single gene, which they dubbed the testis determining factor, or TDF, gene.

This is a good example of an observation that on the surface seems to defy common sense: How could just one little gene be responsible for so many differences between men and women? The proof was provided by an ingenious experiment run by the British scientists who isolated the TDF gene. First, they isolated the corresponding gene from mice, in which sex determination is essentially identical to humans. Next, they injected a small fragment of pure DNA containing the TDF gene into fertilized mouse eggs—male and female—and transplanted these into a surrogate mother.

When the baby mice were born, they were tested for the presence of the transplanted TDF gene. Some of the baby XX mice, which should have been female, now had the TDF gene incorporated into their genome. Remarkably, some of these chromosomal females developed as males, with normal testes and penis. When they matured, they displayed male sexual behavior, seeking out female mice, mounting them and having intercourse. Genetically speaking, these mice were completely female except for one tiny bit of DNA containing the TDF gene, but anatomically, physiologically, and behaviorally, they were clearly male.

The TDF gene plays its key role early in prenatal development, when male and female embryos are still identical. At this early stage, the gonads consist of a mere streak of tissue that will develop into the male or female internal genitalia. The TDF protein, which is synthesized in a small group of supporting cells, does two important things. First, it turns on the gene for a protein called Müllerian-inhibiting hormone, which prevents the formation of the female genital tract. Second, it signals another group of cells to synthesize the male sex hormone, testosterone, which guides the development of the male genitalia and the masculinization of the entire body. Thus male sexual development requires two switches to be activated during a period of only a few days: One switch turns off the female pathway, and the other turns on the male pathway.

In female embryos, which lack the TDF gene and therefore produce no testis determining factor, development follows another path. Since there is no off switch, the embryonic tissue called the Müllerian ducts follows the natural development into the oviducts, uterus, and cervix, while the supporting cells mature into the surrounding tissue for the ova. This process requires no hormones, which aren't even produced in females until after birth.

The hypothesis that sexual orientation in humans might be related to the TDF gene seemed unlikely at first. If homosexuality were something as simple as gay men lacking a Y chromosome, or lesbians having one, it would have been discovered long ago, if not by scientists, then certainly during the cytogenetic sex tests run on Olympic athletes. A more obvious problem with this theory was that if gay men lacked TDF, they wouldn't have penises. There remained the possibility, however, that very slight changes in the structure of the Y chromosome or of the TDF gene itself might cause more subtle alterations in sexual development.

Checking this theory meant looking at chromosome structure, and here we depended on Nan Hu's years of experience as a cytogeneticist. She took the blood samples from a series of people, including gay men, lesbians, gay brothers, and heterosexual family members, and grew their cells in a medium that arrested cell growth just when the chromosomes were most visible. Then she spread the cells onto glass slides and stained them with special dyes to reveal the alternating dark and light bands that characterize each chromosome. The last step was photographing the chromosomes under a high-power microscope and enlarging the photographs to illuminate upwards of 400 distinct bands in each set of chromosomes.

The results from Nan Hu's experiments were clear: There were no differences between the gay and straight men or between lesbians and heterosexual women. Every man had an X and a Y, and every woman had two Xs. These findings were confirmed by directly analyzing the TDF gene coding sequences by the PCR method. As expected, the TDF sequences were present in all the men and absent in all the women, regardless of their sexual orientation.

The "Man-Maker" Hormone

Ask a scientist to free associate on the word "sex," and one of the first responses will be: hormone. For males, the most important hormone is testosterone, which is one of a series of related compounds called the androgens, Greek for "man maker." Once the testes begin to synthesize the chemical testosterone and secrete it into the blood to circulate throughout the male embryo, sexual differentiation is completely controlled by hormones. It makes little or no difference whether the rest of the cells in the body are genetically male or female. As long as they are bathed in testosterone, they will follow the male path of development.

The central role of circulating hormones in sexual development was first deduced in 1849 by A. A. Berthold, a Swedish scientist who found that if he removed the testes from a rooster, it no longer crowed or engaged in male sexual behavior. If he reimplanted one testis in the body cavity, the castrated rooster resumed his normal male activities. Because the transplanted testis did not establish any nerve connections, Berthold concluded that the testes must secrete a chemical that influences the rooster's behavior.

Since Berthold's time, scientists have devised far more sophisticated methods to manipulate hormones in experimental animals, mostly rodents. These experiments have shown that testosterone exerts its influence on three major areas of sexual differentiation: the internal genitalia, the external genitalia, and the brain. They also have shown that the levels of testosterone are not constant, but ebb and flow in a predictable pattern in the womb and after birth.

In humans, the differentiation of the internal male genitalia—the prostate, seminal vesicles and ducts—occurs about two months after conception, during the first burst of testosterone production. The external genitalia—the penis and the scrotum—are formed about four months later, during a period of relatively low testosterone levels. The levels rise again toward the end of pregnancy and continue upwards for the first two months after birth, then drop to low levels. The reason for this birthday surge of hormones is not precisely known. The third and final rise in testosterone occurs at puberty and is responsible for the development of the secondary sex characteristics such as facial hair, a deepening of the voice, and enlargement of the penis.

For sexual orientation, the most interesting role of testosterone appears to be in the brain and the behaviors that it governs. For example, if a newborn male rat is castrated at birth, thereby depriving it of testosterone, it will never display typical adult male behaviors such as mounting and penetration of females. Furthermore, if it is given the "female" hormones estrogen and progesterone during adulthood, the castrated rat will exhibit a typical female behavior called lordosis, which means arching the back to present the rump to an interested male. This effect, however, is very dependent on timing. If the male rat isn't castrated until six days after birth and is primed later with a shot of testosterone, it still will act like a male.

The same kind of gender switch can happen in female rats. When they are exposed to testosterone throughout the critical period immediately after birth, they won't grow up to act like females. If they are dosed with testosterone as adults, they will begin to mount other rats and act like males.

An example of this experimental switching of genders was shown with two rats, known to me as Mork and Mindy, and recorded on video. Mindy was genetically female and dosed with testosterone, while Mork was genetically a male who was deprived of testosterone and injected with female hormones. The X-rated video of their first encounter, made by UCLA researcher Roger Gorski, has a simple and decidedly unromantic plot: Mindy meets Mork, Mindy mounts Mork, Mindy leaves Mork.

When Gorski, who was on sabbatical at our lab, first showed us the tape at "show and tell," I was fascinated. But when I saw it the second time, at a symposium on human sexuality, I was embarrassed. There's no way, I thought, that the stereotypical behavior of rats explains *our* sexuality. Where's the romance? Where's the love? Where's the conscious choice?

Although a scientist in a lab today would never be allowed to experiment on human hormone levels, a few doctors have tried to "cure" gay men with testicular grafts or injections of testosterone. The results should have been predictable: the men didn't stop wanting men, they just wanted more of them more often. Since there is no acceptable way to manipulate hormone levels in humans as in rats, researchers rely on certain natural cases of altered hormone levels, such as genetic conditions that prevent the conversion of testosterone or cause excess production of androgens. There also are many instances in which women have been treated during pregnancy with hormones, such as synthetic estrogens, to prevent miscarriages.

Although scientists have tried to generalize based on these cases of hormone-level change in humans, there is a problem in this approach: In many cases, the body, not just the brain, is changed. For example, in instances where children are born with genitalia that are ambiguous or opposite to what their chromosomes would suggest, they often are surgically "corrected" to female. Naturally enough, these children then are raised as girls, so it's impossible to say if they behave like girls because of hormones acting on their brains or because they were taught to be girls. Clearly, hormones act on the brain; how they act on the brain remains a mystery.

Testosterone's Trigger

The male sex hormone testosterone is a small and simple molecule that directs sexual differentiation by regulating the activity of genes, which in turn produce the proteins that build the physical structures of the body. In some tissues, such as the prostate, testosterone turns on genes that otherwise would be dormant. In other parts of the body, including the brain, testosterone might work by silencing certain genes.

The genes aren't switched on and off directly by testosterone; they need an intermediary called the androgen receptor, which works like a chemical trigger. Since the androgen receptor is central to the differences between men and women, it seemed like a good candidate for something that influences sexual orientation. We started taking a hard look at this in early 1992, when the linkage studies on the gay brothers were just beginning. The gene for the androgen receptor looked promising for several reasons: It already had been mapped to the X chromosome, and more importantly it has two extremely variable DNA sequences.

The presence of these hypervariable sequences was important. Since homosexuality is a relatively common trait, any gene that is significantly involved must have common variations. If the gene differed in only one out of every one thousand people, it couldn't play a role in a trait found in several percent of the population, even if every single person with the variation were gay.

These androgen receptor sequences are not only variable, they also are an interesting type and are located in an important part of the molecule that helps turn on and off the testosterone-regulated genes. This, we thought, might mean variations there influenced the functioning of the androgen receptor in subtle ways. Had the variations occurred in the parts of the molecule involved in hormone binding or DNA binding, it would have been less interesting because these probably would have had a role in visible sexual differences, which are not found between gay and straight men.

The androgen receptor also was a likely place to look because it already had been linked to two important conditions. The first is called testicular feminization, or complete androgen insensitivity syndrome. People with this condition have functioning testes inside their bodies but lack the receptors to use the testosterone. The body is fooled into thinking there are no androgens present, so it develops as female. Chromosomally, the people are males, but they develop as females.

For our theory of sexual orientation, the key point of this gender-switching condition is that people who are chromosomally male (although they look female) are attracted to men. So in one sense, this is proof that a mutation in the androgen receptor gene can "reverse" sexual orientation. This is an overly simplistic interpretation of the evidence, however, since these individuals grow up as women. It actually would be more surprising if they were attracted to women instead of men.

The second genetic condition that has been traced to the androgen receptor gene is called spinal bulbar muscular atrophy, or Kennedy's disease. Kennedy's disease was intriguing for our theory because it is a case where mutations in the androgen receptor gene can alter the gene's function in the central nervous system without necessarily blocking its role in sex determination.

All these factors made the androgen receptor locus an appealing candidate gene for sexual orientation. The only problem, we learned later, was geography: The gene is located far from Xq28, although at this point we had not yet found a linkage there. Our first working theory was that subtle changes in the sequence of the gene would show a correlation with sexual orientation. In other words, people with variations in the gene should have a greater chance of being gay.

To test this hypothesis rigorously, we needed a range of skills and resources unavailable in Bethesda, so we teamed up with four other investigators: molecular geneticist and neuroscientist Jeremy Nathans, population geneticist Terry Brown, and psychiatrist Van King, all from Johns Hopkins Medical School; and Michael Bailey, the psychologist and twin expert from Northwestern University. King, Bailey, and I agreed to provide blood samples from our study groups to Nathans's lab, where his colleague Jennifer Macke would scour them for variations in the androgen receptor gene. Meanwhile, my lab would test our gay brothers for linkage to the same region.

Jeremy Nathans had all the right qualifications for this experiment: He is bright and imaginative, and he had won praise for discovering the genes responsible for color blindness. His father was Dan Nathans, a Nobel laureate. If anyone could prove a relationship between the androgen receptor and sexual orientation, I thought, Jeremy could.

He tested the theory perfectly. Over the next year, Jeremy's lab checked DNA samples from 197 homosexual men against DNA samples from 213 people from an earlier vision study, most of whom were assumed to be heterosexual. After several months and a few thousand assays, the results were in: There was no difference between the androgen receptors in gays and straights. Although Jeremy and Jennifer did discover a few new mutations in the androgen receptor coding sequences, none of them correlated with sexual orientation. Likewise, Stella Hu's linkage study showed that there was no association between the inheritance of the androgen receptor gene and sexuality in our group of gay brothers.

Even though the findings were negative, we were able to draw two important lessons from the androgen receptor experiments. First, we don't really know enough about human sexuality to have a good feel for which genes might be important in the development of sexual orientation. We can make educated guesses, but they're still guesses.

Second, just because a particular gene or protein is required for a particular developmental pathway, it doesn't necessarily follow that variations in that pathway are caused by variations in that particular gene or protein. In our case, this meant that although the androgen receptor is on the pathway to both heterosexuality and homosexuality, it's not at a fork in the pathway. A simple illustration of this idea is that while the protein keratin is necessary to grow hair, it doesn't follow that variations in the keratin gene will make some people blond and some brunette.

This was another reminder that genetics is the study of variation. With the tools available so far, the science can only identify those genes that contribute to differences in a trait, not all the genes required for a trait everyone shares.

Lefties

"Batting lefty" is a quaint euphemism for being gay. Some researchers believe it's partly based on truth, that there is in fact a right to left laterality shift in homosexuals that signals a difference in the hormonal control of brain development.

In the general population, approximately 90 percent of people are right-handed and 10 percent are left-handed, as determined by the question, "Which hand do you write with?" The more sophisticated Annett handedness questionnaire, which includes items such as "Which hand do you use on top of a push broom?" has shown that about 30 percent of the righties actually have a "nonconsistent right-hand preference," meaning they write with their right hand, but they have some degree of ambidexterity.

The idea that there might be a connection between hand preference and gonadal steroids has been broached by several researchers, but the evidence is nearly as controversial as that linking hormones to sexual orientation. Nevertheless, given how easy the experiments are to perform, it was inevitable that researchers began to look for associations between which hand people favor and their sexuality.

We were sufficiently intrigued by these published studies to ask all our subjects, both gay and straight, about their hand preference. We used a version of the Annett questionnaire that included those items that give the highest predictive estimates of composite handedness. The final results were clear: The gay men showed the same fraction of left-handers and nonconsistent right-handers as did the heterosexual men, and both groups were indistinguishable from the general population. The data for the lesbians had not been completed in early 1994.

Despite this negative finding, there was one interesting observation. Among pairs of gay brothers, 43 percent of the seven Xq28-discordant pairs compared with only 6 percent of the thirty-three Xq28-concordant pairs were discordant for handedness; that is, one brother was right-handed and the other was left-handed. Because of the small number of pairs involved, the difference was too small to be statistically significant, but it does raise the interesting question of whether some men are gay because of genes and others because of differences in brain development that correlate with handedness.

The Hypothalamic Hypothesis

If our most important sexual organ is the brain, perhaps the Xq28 locus influences sexual orientation by altering the brain's structure or chemical activity. For this to be a plausible hypothesis, it first would have to be demonstrated that there are detectable differences between the brains of heterosexuals and homosexuals.

Simon LeVay, a neuroanatomist, is convinced that there are such differences, and that he's found one of them. In 1991, LeVay, who then was the director of the vision laboratory at the Salk Institute in San Diego, published an important paper in *Science* called "A Difference in Hypothalamic Structure Between Heterosexual and Homosexual Men."

LeVay's study examined the brains of the cadavers of gay men, heterosexual and presumably heterosexual men, and presumably heterosexual women. He focused on the hypothalamus, a small area at the base of the brain that plays a key role in our most basic urges: hunger, thirst, and sex. LeVay cut the hypothalamus into paper-thin slices, mounted these on slides, stained the neurons with dye, and examined the tissue under a microscope. This was done "blind," using identification codes, so LeVay wouldn't know if he was looking at a brain from a gay or a straight individual.

LeVay was especially interested in a small group of cells called the third interstitial nucleus of the anterior hypothalamus, or INAH-3. Two years earlier, Laura Allen, a postdoctoral fellow in Roger Gorski's lab at UCLA, had shown that this structure was significantly larger in men than in women. (The sexual orientation of the subjects was unknown, but presumably most were heterosexual.) LeVay hypothesized that there would be a similar difference, or dimorphism, related to sexual orientation; that is, that INAH-3 would be larger in heterosexual than in homosexual men.

To test this, LeVay measured the volume of INAH-3 in his brain preparations. First he replicated the experiment of Allen and her colleagues and found INAH-3 on average two or three times greater in volume in the presumably heterosexual men than in the women. The second result broke new ground: INAH-3 in the gay men was the same volume as in the women, or two to three times smaller than in the straight men. Although there was a great deal of overlap between the homosexual and heterosexual men, a statistical test indicated that the difference was significant. Critics argued that the twofold difference in cell size was too small to matter, but a similar doubling of the entire organism would produce a 12-foot, 350-pound person.

The critics also pounced on the fact that all the gay men in LeVay's study had died of AIDS, so perhaps the size of their INAH-3 was related to viral infection or its treatment rather than to sexual orientation. LeVay, of course, was aware of that possibility and compared his gay subjects with heterosexuals who had died of AIDS and found the heterosexuals still had the larger INAH-3. In a separate study after the paper was published, he found that a gay man who had died of lung cancer had the smaller INAH-3 found in the gay men who had died of AIDS. As a third control, LeVay examined three other groups of cells in the hypothalamus and found no differences between the gay and straight men.

LeVay's cautious conclusion was: "This finding . . . suggests that sexual orientation has a biological substrate." The claim is hardly radical, because neuroscientists consider most mental processes—whether choosing a sexual partner or deciding what tie to wear—ultimately to be biological. What was surprising was that LeVay found this mental "substrate" localized within one small region of the brain and that the differences were so marked.

INAH-3, together with several other nuclei and axonal tracts, is located toward the front of the hypothalamus in a region called the preoptic area. Experiments involving stimulation or surgical removal of this area suggest it is intimately tied to male sexual behavior, and even sexual "feelings," in animals. In an experiment cited by LeVay in his book *The Sexual Brain,* a male monkey sits calmly by a female monkey in heat. When the male's preoptic area is tingled with a weak electrical current, he promptly mounts the female. This apparently isn't just a motor reflex, because tingling won't get him out of his chair if the female isn't in heat.

Another experiment placed recording electrodes in the preoptic area of a male monkey who was strapped to a chair. A female monkey in heat was placed in another chair within view. By pressing a button, the male monkey could bring the female close enough for (a presumably rather cramped version of) intercourse. The recordings showed that pressing the button caused neurons in the male monkey's preoptic area to fire at a high rate. When a banana was placed in the movable chair instead of a female monkey, the male was equally eager to press the button, but there was no change in the electrical activity of the preoptic area.

In rats, this part of the brain contains a grove of cells that is about fivefold larger in males than in females. Roger Gorski and his colleagues at UCLA have shown that the growth of this nucleus is regulated by testicular hormones and correlates with male sexual behavior. They called the area the sexually dimorphic nucleus, or SDN, and found that its size was reduced by castration and restored by hormone injections. In general, the bigger the SDN, the more typically masculine behavior is. In recent, still unpublished experiments, Gorski and colleagues found that electrical stimulation of the SDN, but not of the nearby regions, caused male rats to display a variety of hypersexual activities. The relevance of this to LeVay's work is that the anatomical position of the rat SDN is very similar to the position of INAH-3 in the human brain. More study is needed, however, to see how similar SDN and INAH-3 really are.

The evidence he found convinced LeVay that INAH-3 might be intimately involved in human male sexual behavior; not that it "makes" male sexual feelings, but that it plays some role in triggering or transmitting sexual impulses in the brain. One problem with such neuroanatomical studies, however, is they can never prove causality: Do differences in the size of INAH-3 cause homosexuality, or is it the other way around?

Two observations suggested to LeVay that the structural differences in INAH-3 might precede rather than follow sexual orientation. First, the presumably corresponding region of the rat brain is sexually differentiated during the first five days of life, well before sexual maturation at puberty. Second, INAH-3 volume is not the only observed difference between the brains of gay and straight men. Recently Allen and Gorski found another structure, the anterior commissure, that was larger in gay men and in women than in presumably heterosexual men. Scientists in Holland came up with still another region, the suprachiasmatic nucleus, which was found to be larger in homosexual men than in either straight males or heterosexual women. Neither of these two structures is known to have any direct role in sexuality, however, so it is unlikely that the size differences were the result of differences in sexual behavior. Rather, they may signal an early developmental difference in the brains of straight and gay men that can influence a number of different brain structures and circuits.

LeVay's paper in *Science* began the modern era of scientific research into sexual orientation and attracted a flurry of publicity and commentary. Despite the numerous criticisms of his work, no one has tried to replicate the experiment, which should be the first order of business. Data speak louder than words, but so far the labs are silent.

Where might a "gay gene" fit into LeVay's analysis? The most simple hypothesis would be that Xq28 makes a protein that is directly involved in the growth or death of neurons in INAH-3. Alternatively, the gene could encode a protein that influences the regulation of this region by hormones. Only the isolation of the Xq28 gene, and further testing of LeVay's hypothesis, will provide the answer.

Chapter Ten

PSYCHOLOGICAL MECHANISMS: SISSIES, FREUD, AND SEX ACTS

The idea persists that there is a division between the brain—the organ in the head—and the mind, or consciousness. In science, this translates into the division between biology and psychology, and despite new disciplines called psychoneurobiology and biological psychology that attempt to bridge the gap, biologists and psychologists still have different ways of thinking about thought. Biologists are interested in seeing if a gene related to sexuality might shape or influence physical structures or biological mechanisms—the testis determining factor gene, the androgen receptor gene, or a specific part of the brain. Psychologists are more inclined to look at how such a gene might be involved in personality and psychological mechanisms that influence who is gay, who is straight, and who is in between.

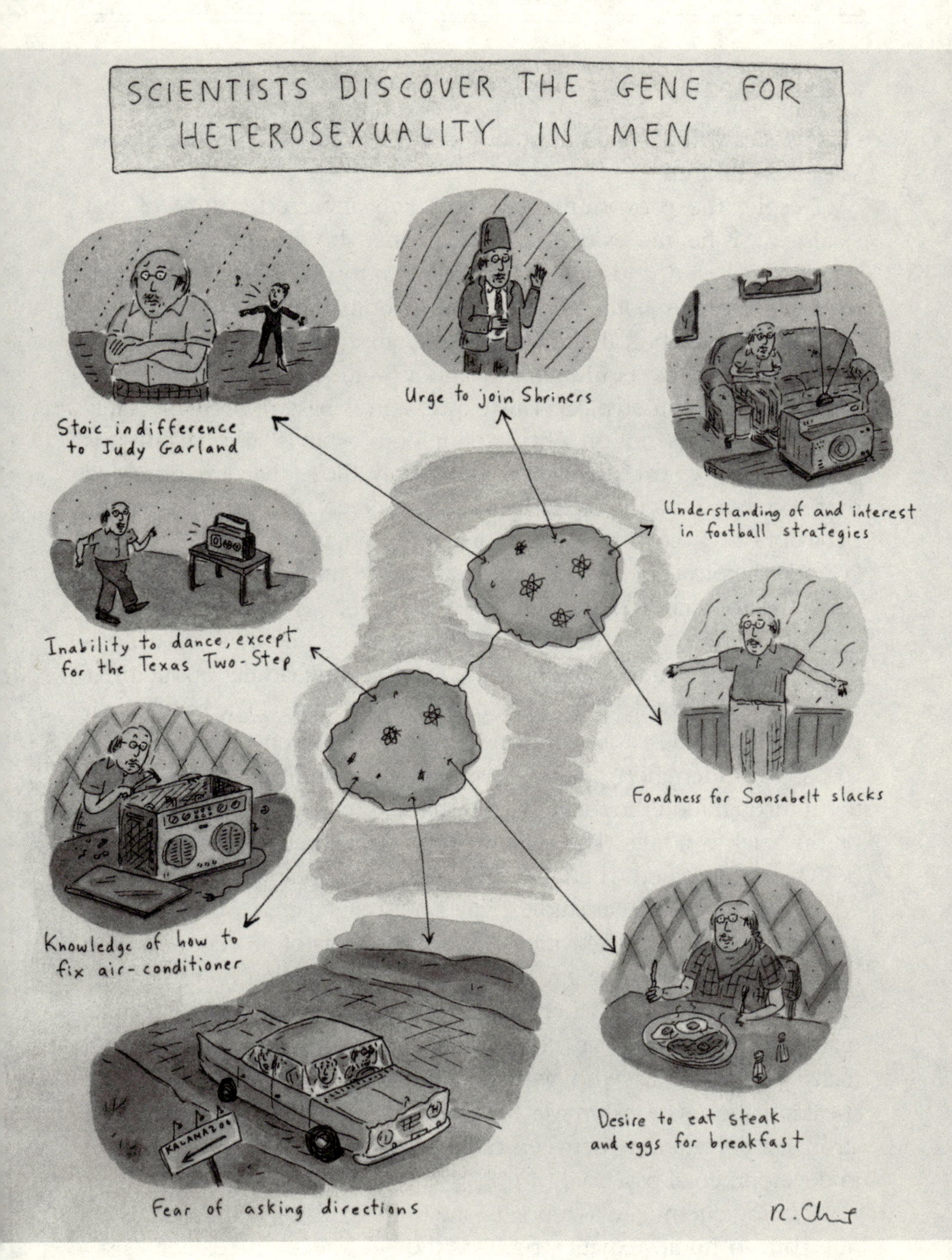

Drawing by R. Chast; © 1993
The New Yorker Magazine, Inc.

The "Sissy Gene"

Most sissies will grow up to be homosexuals, and most gay men were sissies as children.

Despite the provocative and politically incorrect nature of that statement, it fits the evidence. In fact, it may be the most consistent, well-documented, and significant finding in the entire field of sexual-orientation research and perhaps in all of human psychology.

The more technical and scientifically appropriate term for "sissy" —and the female equivalent, "tomboy"—in discussing childhood behavior is gender-atypical. This expression is based purely on statistical observations, not on a firm definition of what is "normal." In this sense, "male-typical" behaviors are simply those that are observed more often in boys than in girls in a given society, and "female-typical" behaviors are those that are more common in girls than in boys. "Gender-atypical" behaviors are those that are unexpected, on an average basis, for a child of the gender in question.

The connection between childhood behavior and adult sexual orientation has been documented by interviewing adults about their childhood and by tracking children as they grew older. The best known prospective studies were conducted by Richard Green, a psychiatrist at UCLA, who identified notably "sissy" boys as children and followed them through adulthood. He found that the majority identified as gay or bisexual as adults. The study is remarkable because some of the boys were identified when they were only a few years old.

In the retrospective studies, gay and straight men were asked to remember childhood activities. A recent analysis of 32 reports in the literature found that the gay men consistently recalled more gender-atypical behaviors than did the heterosexuals. There were, of course, many gay men who had conventionally masculine childhoods and many heterosexuals who were effeminate as boys, but statistically speaking, the relationship was highly significant. In fact, it is one of the strongest correlations between childhood and adult behavior known in developmental psychology.

If being effeminate as a boy is one of the consistent milestones on one path to homosexuality, perhaps the "gay gene" is really a "sissy gene." To find out, we reviewed what our gay and straight participants had told us about their childhood during interviews using the standard questions for this type of research. For example, "Did you consider yourself less masculine than other boys your age, or were you ever regarded as a sissy as a child?" The answer was yes for 68 percent of the gay men, compared with 5 percent of the straight men. Another question was, "Did you enjoy sports such as baseball and football as a child?" Of the heterosexual men, 78 percent said "very much," compared with only 8 percent of the homosexual subjects.

When we tallied up the responses to all the questions, we got the expected result: The gay participants recalled substantially more gender-atypical behaviors than the straight subjects. This probably was not a result of gay men being more willing or able to remember acting like "sissies," because in several of the families the recollections were confirmed by parents or siblings.

This information allowed us to test the "sissy gene hypothesis" by seeing whether Xq28 harbored a gene that correlated with gender-atypical behavior. If it did, brothers who shared this part of the chromosome would be more gender atypical than the brothers who were discordant for this region. But when we compared the two groups, the evidence did not support the hypothesis. If anything, the opposite was true: The Xq28-concordant pairs recalled slightly more typically male behavior than the discordant pairs.

Another way to look for possible connections between genes, atypical behavior, and sexual orientation was suggested by Michael Bailey. He compared the levels of gender-atypical childhood behavior in gay men who had gay identical twins to those who had straight identical twins. The two categories of men had indistinguishable scores, indicating that the factors responsible for homosexuality and gender atypicality did not cause each other. Using a similar approach, we compared gay men with gay brothers to gay men without gay brothers and also found similar levels of gender atypicality.

Despite these findings, there are some intriguing observations that suggest there may yet be a connection between the "gay gene" and gender-atypical behavior. In his twin studies, Bailey noted that gay-gay pairs had more similar gender-atypicality scores than did gay-straight pairs. Similarly, we found that Xq28-concordant gay brothers were more alike in their childhood behavior than Xq28-discordant pairs. Perhaps the Xq28 locus somehow modifies or interacts with the factors responsible for gender-atypical behavior.

The "Bottoms-Up Gene"

Being a "sissy" is not limited to boyhood. Some adult gay men also are regarded as "swish," "camp," and effeminate, but we made no attempt to measure gender atypicality in adulthood because other research had shown that, in reality, it correlates quite poorly with sexual orientation. It also seemed intuitively obvious that this dimension of personality is too much influenced by the social and cultural environment to have any significant biological or genetic component.

There is, however, one aspect of adult gay male behavior that is notably and measurably gender atypical, and that is the sex act itself. The very fact of having sex with another man is in and of itself gender atypical, and of the large number of different things a man can do sexually, the most gender atypical is to take the receptive role in anal intercourse.

Every subject in our study was questioned about preferred and performed sex acts. This was important not just to understand their sexuality, but also to learn about their risk for HIV infection, information we needed for the study of AIDS. Both the gay and straight participants were asked about the following sex acts: touching and masturbation, oral-genital, penile-vaginal, penile-anal, oral-anal, and "other." As appropriate, the subjects were asked whether they engaged in these activities insertively, receptively, or mutually. They also were asked what activities they would prefer to perform, or fantasized about, as compared to what they actually did. This was important to see if people had altered their behavior because of the risk of AIDS and other sexually transmitted diseases.

Among the heterosexual men we interviewed, the insertive role in penile-vaginal intercourse was by far the leading choice in fact and fantasy, and the range of other activities was limited. The gay men exhibited a more flexible and diverse repertoire of sexual activities and desires. Mutual oral-genital contact and masturbation were about tied for first place, penile-anal intercourse was next, and oral-anal and "other" were last. Only 16 percent of the men listed receptive anal intercourse as their preferred sexual activity, either in reality or in fantasy.

This information allowed us to test the hypothesis that the "gay gene" is actually a locus for receptive anal intercourse. Using the same strategy we used for the "sissy gene," we divided the 40 pairs of gay brothers into Xq28-concordant and Xq28-discordant groups and compared the number of men who listed receptive anal intercourse as their preferred sexual activity. There was no significant correlation of any sort, nor was there a difference among all gay men with gay brothers compared to those without gay brothers. In fact, there was little correlation at all between the sexual activities preferred and engaged in by gay brothers, indicating there is as yet no evidence that the particular sexual activity a person prefers is genetically influenced.

The "Self-Sufficiency Gene"

Another model of how genetics influences sexual orientation is that Xq28 contains a gene for the personality trait called self-sufficiency, which in this sense means iconoclastic, or socially independent. The hypothesis is based on four propositions: Self-sufficiency is a prerequisite for homosexuality in the population we studied; there are differences between gay and straight men in their degree of self-sufficiency; self-sufficiency is genetically influenced; and one of the genes for this characteristic is located in Xq28. As unlikely as it sounds, there is some evidence gathered by other laboratories to support these claims.

The psychologist Raymond Cattell describes self-sufficient individuals in the following terms: ". . . temperamentally independent, accustomed to going their own way, making decisions and taking action on their own . . . they discount public opinion . . ." The opposite of self-sufficiency is "group reliance," which describes "joiners," who depend on social approval and admiration for self-esteem. Cattell measured these traits with what he calls a sixteen-personality-factor inventory, or 16PF. He got data for the scale by asking subjects to respond to statements such as "Most people would be happier if they lived more like their friends and did much the same thing as others."

Based on this specific definition, our subjects—all openly gay men —must have been self-sufficient. They not only had to be sexually attracted to other males, they had to ignore society's scorn for homosexuals and acknowledge their orientation to themselves, to friends and family members, and to an outside researcher.

The second proposition, that self-sufficiency is correlated to sexual orientation, has been tested by administering the personality inventory to gay and straight men. In each of three studies conducted in Australia, the United States, and South Africa, gay men scored considerably higher than straight men for self-sufficiency. This "positive" trait was one of the most significant differences between gays and straights in two of the studies, and even in a 1962 study for the "Australian Commission to Study the Problem of Homosexuality," which concluded that homosexuals have a "criminal" personality profile.

The role of genetics in self-sufficiency and other personality traits has been approached primarily by comparing identical and fraternal twins. The main conclusion is that virtually all personality traits, including those that contribute to self-sufficiency, are to some extent influenced by genes. The usual estimates are that genes contribute between 20 percent and 60 percent to personality, while other factors, such as family life, explain the rest. Although the methodology of these studies has been criticized, the accumulated evidence indicates that personality traits are at least somewhat influenced by genes.

The final proposition—that self-sufficiency is specifically influenced by a gene at Xq28—was based on a tantalizing but still inconclusive experiment conducted by Jonathan Benjamin, R. H. Belmaker, and their colleagues at Be'er-Sheva Mental Health Center in Israel. Benjamin, who conducts psychiatric screening tests for the Israeli army, identified 17 pairs of brothers in which at least one brother was color blind. Because the color-vision locus is located in Xq28, it allowed him to determine which brothers shared DNA sequences in this region of the X chromosome.

Benjamin administered the Cattell 16PF test to all the brothers and tallied up the scores on each of the sixteen items. If a gene in Xq28 influenced one of the personality traits, he expected the Xq28-concordant brothers (both color blind) to be more similar to one another than the pairs with only one color-blind brother. This pattern was observed for only one area of the test: self-sufficiency.

When I received Benjamin's manuscript, a few months after our *Science* paper was published, I was astonished. I never would have thought of looking for a "self-sufficiency gene." Unfortunately, when I calculated a LOD score, it was only 0.8, which was 160 times lower than the usual cutoff of LOD 3.0, and 1,600 times lower than the LOD score of 4.0 that we reported for sexual orientation. But despite the poor statistics and inconclusive results, I thought the theory deserved a closer look.

In early 1994, we were sending out the 16PF questionnaire to all 40 pairs of gay brothers in our original study and administering it to all new volunteers. We also were collaborating with Jonathan Benjamin to replicate his experiment, using a much larger group of heterosexual brothers and using real DNA markers instead of the "poor man's marker" of color blindness. This was a great deal of effort to base on a hunch, but my own interviews for our study had turned up some interesting tidbits that now were beginning to make sense. For example, there was the case of the brothers I'll call George and Albert.

George answered the phone on a summer day in 1992 and gave me directions to the house he shared with his brother. He added, as a sort of caution, that his brother was "kind of conservative." I followed the directions and pulled up in front of a 1950s ranch-style house in a quiet, residential neighborhood in a small, southern hamlet. The grass was green and well tended, and pink roses swayed alongside the white picket fence.

I rang the bell and was greeted by a man who was naked from the waist up. Two shiny metal rings hung heavily from his pierced nipples, and there were thinner rings through his lip and eyebrow. His torso was busily tattooed, and when he turned to lead me into the living room, I noticed that they extended across his back and included a German cross and a wreath of snakes and bullwhips.

"You must be the less-conservative brother," I ventured.

This was George, and he was not a conventional guy. At age 13, he said, he seduced his pastor. At 18, he left home for New York to begin a career as a stripper, a hustler, and a gigolo. When I asked him to estimate his lifetime total of sexual partners, he said that he had thought quite a lot about the number. Including every contact at bathhouses, sex clubs, and backroom bars, he figured it was 10,000 to 20,000 partners, including several hundred women he'd slept with for money.

When I ran the standard questionnaire on him, including demographics such as educational level, I was impressed to hear he had gotten a master's in New York.

"Not a master's," he corrected me. "A master. I've got a master in New York."

Even with a history of sadomasochism, extensive drug use, and multiple partners, George tested negative for the AIDS virus. If there is a gene that resists the AIDS virus, I thought, George would be a good place to start the search.

When Albert appeared, I thought he was George's accountant, or maybe the piano teacher, rather than his only brother. Prim and proper, Albert said he had a 9-to-5 job, went to church on Sundays, and was the one responsible for keeping the yard so neat.

The one thing Albert shared with his brother was an attraction to men, which he first recognized at age 5. For the next twenty-two years, however, he did everything possible to deny those feelings to himself and to others. When he fell in love with a college classmate and was rejected, he sought solace in becoming a Jehovah's Witness. When his homosexual attractions still wouldn't go away, he joined the marines to prove he was a real man. Only after an emotional breakdown, which led to counseling from a sympathetic psychiatrist, did Albert begin to accept his sexuality. As a gay man, he still struggled to lead a conventional life, and following the breakup of a six-year relationship, he remained chaste because he preferred a monogamous relationship.

According to the definitions used in the personality test, Albert was group oriented, while George was the very picture of self-sufficiency: insensitive to, if not downright disdainful of, social norms. When we tested their DNA, we found that Albert and George not only had different personalities, they had different versions of Xq28.

Since both men were gay but had different versions of Xq28, perhaps the gene was influencing something related to their psychological adjustment to their homosexual orientation rather than to their sexuality per se. For example, perhaps during childhood they both developed gay fantasies, for whatever reason, but only George had a version of Xq28 that led him to indulge those fantasies. Albert, on the other hand, had different DNA and suppressed his gay fantasies to be more "normal."

This analysis is purely a guess, but it suggests an interesting way of looking at how an inherited personality trait unrelated to sex might influence sexual orientation.

BACK TO THE FUTURE

We have never suffered a shortage of psychological theories to explain homosexuality or other complex behaviors. Facts to support the theories, however, have been in short supply. At the turn of the century the belief was strong that people were innately different and that the human race could be "improved" through selective breeding. After World War II and Hitler's attempts to build a "master race," new theories emerged that placed more emphasis on the rearing environment than on inborn traits. Now the pendulum seems to be swinging back toward the influence of genes on personality, without discounting the role of parents and society in shaping a child. Although many of the early theories about sexual orientation have fallen under the weight of new empirical evidence, some have become myths that continue to sway the public perception of homosexuality.

Freudian Fantasies

By far the most influential thinker about the origins of homosexuality was the Austrian physician and neurologist Sigmund Freud, whose turn-of-the-century ideas form the foundation for many popular perceptions of human sexuality. Freud himself had, at different times, different and contradictory ideas about homosexuality. Early on, he seemed to think that homosexuality might be a nonpathological variation of behavior with innate underpinnings. Later, however, he came to see homosexuality as a pathological state of arrested psychological development caused by inadequate or inappropriate parenting.

One of Freud's theories was that homosexuality is the result of a "castration complex," which a man develops when he realizes that his mother doesn't have a penis. Fear of losing his own penis drives the man to have sex with male partners instead of women. An even more influential theory of Freud's was that homosexuality is caused by a failure to break the sexual bond with a smothering or dominant mother. This supposedly is abetted by a weak or absent father who fails to provide a model of "normal" male development. Later in life, the failure to separate from the mother manifests itself as a desire to "be" the mother in sexual relationships.

Freud's ideas were popular and soon dominated both the psychiatric profession and public perception. Perhaps this was because his theories were unencumbered with any bothersome facts, perhaps because they made liberal use of metaphor and myth, or perhaps because they allowed people to blame their "condition" on others. Freud's theories gave scientific credence to the idea that gay people were sick, that they could change, and that they'd be happier if they did.

There are at least two central deficiencies in Freud's analysis. The first is the failure to separate cause and effect. As the psychoanalyst Richard Isay has pointed out, it might not be the parents who cause their children to be gay but the gay children who cause their parents to react. Perhaps prehomosexual boys show early signs of being gay that cause their mothers to become overly protective and their fathers to become distant. To support this idea, sociologists have shown that gay men brought up in societies in which homosexuality is tolerated have less hostile relationships with their fathers than those raised in intolerant societies.

The second problem with Freud's theory is that it doesn't fit the facts, at least not in the families interviewed for our study. Most psychoanalytic literature is based exclusively on patient self-reports, and analysts reconstruct family histories based on what they are told by a single family member, the patient. In contrast, we had the opportunity in some cases to interview entire families—not just the gay subjects but also their siblings and parents.

Not surprisingly, most of the gay men we interviewed remembered being closer to their mothers than to their fathers—but so did their heterosexual brothers. In a few families, the gay men were closer to their fathers—but in those families, so were their straight brothers and sisters. Furthermore, all the parents interviewed were quite certain they'd treated their children, whether gay or straight, very much the same. These observations seem to contradict Freud's description of homosexuality as an outgrowth of distorted family dynamics.

This isn't to say that parental relationships aren't tremendously important for a person's psychological and sexual development, for it's within the family that a person's personality and sexuality unfold. Nor is it to say that modern psychiatry can't help people deal with psychological ups and downs. The problem lies in extrapolating from these theories to say that the mind is a blank slate upon which only the parents can write.

Behaviorism

During the past thirty years or so, serious students of sexuality largely have abandoned the ideas of Freud. They have not, however, given up on the idea of the mind as a tabula rasa, a blank slate upon which experience is inscribed. The new behaviorists have simply changed their ideas about who does the writing—from dominant mothers and absent fathers to peers, teachers, and scout leaders, or to society and culture in general.

The behaviorist school of thought, of which Kinsey was a member, holds that adult sexual preferences are shaped largely by early erotic and social experiences. For example, if a boy's first sexual experience is a pleasurable contact with another male, he might grow up to be homosexual, but if he has an enjoyable early experience with a girl, he would more likely develop into a heterosexual. The problem with this idea, like Freud's theory before it, is that it doesn't separate cause and effect. If the little boy prefers other boys in the first place, he naturally would be more likely to seek them out for sex.

Similarly, the behaviorists say, if the boy is exposed to homosexual role models early on, he might be tilted toward being gay. The trouble with this notion is that it doesn't fit with the evidence that later was collected by Kinsey's own institute. These later studies showed gay and straight men had no significant differences in their childhood contacts with homosexuals.

A variation on this theme is called social learning, which holds that both heterosexual and homosexual development are the result of societal expectation; that people are gay or straight because they are taught to be that way. This notion was most clearly articulated by C. A. Tripp in an influential book called *The Homosexual Matrix,* in which he wrote: "Most people see their heterosexual responses as innate and automatic, but trained observers understand that people are specifically heterosexual because they have been geared by their upbringing to expect and want to be."

There are many problems with this extremist view of the role of society in shaping individuals. First, it predicts that in societies in which homosexuality is tolerated or encouraged, the numbers of gays will be increased or even predominate. Yet even among the Sambians, a New Guinean tribe in which young boys are told they must fellate older men and drink the "father's milk" (semen) to grow, adult homosexuality is the exception rather than the rule. More mysterious still is why there are any homosexuals at all in most societies, since any signs of homosexuality are actively discouraged.

A second problem with the social learning theory is that it doesn't make any biological or evolutionary sense. We would have to believe that during the course of evolution, all our "animal" instincts for courtship and mating suddenly disappeared; drives such as hunger and thirst survived, but sexuality was suppressed. Socialization, learning, and experience certainly influence the types of sexual and romantic activities people enjoy, but these social factors can't possibly account for the survival of the human race.

The trendiest of the behaviorist schools is called social constructionism. It postulates that there is no such thing as heterosexuality or homosexuality, only definitions of sexuality that are imposed by culture. Proponents of this theory are fond of pointing out, at every opportunity, that prior to 1892 the word "homosexuality" did not even exist in English. Therefore, they argue, homosexuality (without the quotes) is merely a cultural label; it has no universal meaning, much less any biological component.

This kind of thinking in other areas would have left us in the dark, literally. There wasn't really a unified theory of "electricity" until James Clerk Maxwell came up with one in 1864, and electricity has been well understood and harnessed only during the past one hundred years. So, taking social constructionism to the extreme, electricity did not "exist" until quite recently and even now has no "real" meaning. Try that theory on someone who's been struck by lightning.

Ironically, the strongest advocates of social constructionism in sexuality are the *Journal of Homosexuality* and certain academic gay studies centers, institutions that, if their theories about the unreality of homosexuality were correct, should not even exist. Fortunately for them, the social constructionist theory is not likely to be disproved any time soon, since its content is too amorphous to ever be tested rigorously.

Genes Redux

It's easy to say that nature and nurture work together and play important roles in human sexual development, but the tricky part is determining what the biological and psychological factors are, how they interact, and how much each contributes to the individual variations in human sexuality.

One way to think about the forces that mold sexual orientation—or any human characteristic—is with a matrix. Along the vertical axis of the matrix, write all the possible genotypes of a person. Along the horizontal axis, list all the possible environments and experiences that a person could encounter now, in the past, or in the future. That number is quite large if not infinite, and it certainly is as large as the number of genetic variations. Then write in each little square of the matrix whether the occupant is gay, straight, or in between.

The next step would be to determine the average Kinsey score, or some other measure of sexuality, for the entire matrix and to calculate the amount that each person deviates from that average score. To determine the genetic contribution, average the variances down each vertical column. To get the nongenetic contribution, repeat the same procedure along the horizontal rows. This will produce two numbers that tidily summarize how much of sexual orientation is inherited and how much is environmental.

The only problem with our matrix is that the number of little squares formed by the two axes would be larger than the number of people who have ever existed on the Earth or who will be born before the sun explodes and our planet disappears. That's why researchers spend their time trying to reduce the numbers of columns and rows and puzzling over the invariably incomplete results.

A similar matrix drawn by Gregor Mendel for one of his pea-plant experiments would have had just two rows for genotypes—either a plant has the gene for wrinkled seeds or the one for smooth seeds—and only a single column for environment—the monastery garden. This rightly would be called a genetically determined matrix, and Mendel would be labeled a biological determinist.

A matrix for human sexuality drawn by a classical psychiatrist or psychologist who discounts the role of genes would have only a single row but a large number of environmental columns. A social constructionist might draw something with a single row and a single column but would expect people to view the matrix through a million different pairs of glasses.

My own matrix would have billions and billions of rows, but only the first few would be labeled with the relevant Xq28 genotypes. The rest would also be question marks, at least for now. There would also be plenty of columns, but the only ones that could be filled in at this point would be the ones for experiences and environments typical of present-day, middle-class American men. Many other columns, such as those labeled ancient Greece or modern Brazil, would also be question marks. Thus most of the matrix would be blank, awaiting fresh facts and discoveries.

If the matrix analogy is too mathematical, try another. This book was written on a computer. But what is "the computer"? One component is the hardware—the chip, the electrical connections, the plastic case, etc. A second component is the software—the DOS operating system and programs for word processing and statistics. The third component is the data files, including this text. Which is most important? They all are, of course; any one component would be useless without the others and even to think about them separately makes no sense.

Genes are hardware. They make a person human, produce a penis or a vagina, and allow sexual activity. Data files are the life experiences and environment: the first date, the first kiss, sexual encounters. The software is the difficult part to translate. The data of life's experiences are processed through the sexual software into the circuits of sexual identity. I suspect that the sexual software is a mixture of both genes and environment, in much the same way that the software of a computer is a mixture of what's installed at the factory and what's added by the user. An IBM operating under DOS can have many varieties, but it will never behave like a Mac. Likewise, some men probably are born with genes that ensure they will be sexually attracted to women—and some to men—all of their lives.

The choice of a word-processing program influences how a person will write on the computer, but the decision about which program to use might be based on chance. Likewise, a person's sexuality is set up by the chance inheritance of a particular set of genes and is swayed by personal history, society, and culture. It's the intermingling of nature and nurture that makes each person unique, and trying to understand a single element in isolation is like trying to read a computer diskette by holding it up to the sun.

Chapter Eleven

EVOLUTION

Genes are selfish. They only think about themselves. For an individual gene, the human body is just a temporary vessel to be used briefly and discarded on the march through time. The gene has only one mission—to endure—and the only way it can continue to exist is if its host multiplies and passes on the genetic information to the next generation. The cold, calculated process of evolution is mercilessly unkind to genes that don't contribute to reproduction, cleansing these genes from the species, causing them to die out quickly.

That's why it would seem almost impossible for there to be a "gay gene." After all, how could a gene that discourages reproduction survive more than a few generations? Imagine, just for the sake of argument, that sexual orientation in men were controlled by a single "gay-straight" gene on the X chromosome and that every man who inherited the gay version of the gene was gay and had no children, while women with the gay version were heterosexual and had the usual number of children. Even if half the male population had the gay

"If homosexuality is inherited, shouldn't it have died out by now?"

version to start with, the gene soon would die out because the men would have no offspring. The women would pass the gene to half their children, but the frequency of the gene would decline with every generation, like spilling half the water each time it's poured from glass to glass. In five generations, only 2 percent of the men would be gay, and in twenty generations—barely an instant in evolutionary history—only one out of a million men would be gay.

We know gay men do in fact have children, but as long as they have fewer children on average than heterosexuals, the end result would be the same. The gene would survive for more generations, but it still would disappear. These are mathematical facts: If at some point in evolutionary time, a "straight gene" had changed into a "gay gene," and if every man who had the gene was exclusively homosexual, the gene never would have spread in the population. Even if this type of a "gay gene" somehow appeared in a percentage of the population today, it would be gone tomorrow.

It would appear something must be wrong with this simplistic theory of evolution or with our finding of a genetic link to homosexuality. But, in fact, there are several ways that a "gay gene" could survive, even flourish in the population. What if, for example, a "gay gene" actually increased rather than decreased the number of offspring in some of the people who carried it? Or what if the gene existed in an unusually unstable part of the genome and was continually changing during the rough-and-tumble passage of generations?

At this point, the only explanations for the evolution of "gay genes" are purely theoretical, but the various theories might help to explain how a gene for sexual orientation works in the real world.

TIME LAG

Genes are enduring, not fashionable. Most of our genes were selected in a long-ago, preagricultural era, the Pleistocene. Perhaps under those harsh conditions people who had the "gay gene" weren't even gay, or if they were, they still had the same number of children as everyone else.

Even in modern times, there probably have been substantial changes in the relative reproductive rates of gay men. Our study found that the openly gay young to middle-aged men we interviewed had approximately one tenth as many children as their heterosexual brothers, but a study of San Francisco–area men, performed twenty years ago by Bell and Weinberg, found that the gay men had one fifth as many children as their heterosexual reference group. It's not hard to imagine that fifty or one hundred years ago, when social conditions were far more restrictive than today, many more gay men remained in the closet, married women, and had children.

This line of thinking leads to an interesting speculation. Restrictive societies, in which gay men are likely to have children, should have the most "gay genes." The flip side is that "permissive" societies that allow gays to express themselves and marry other gay men might be the first societies to witness the extinction of gays, at least the ones whose homosexuality was influenced by genes.

FERTILE FEMALES

Although the time-lag model might explain how a "gay gene" evolved in the past, it doesn't explain how it survives at present. One possibility is that the gene increases the reproductive rates of women who carry it. Our family studies showed that most of the female relatives of gay men are heterosexual. If these women had *more* children than average, the additional births would compensate for the loss of the gene through the gay men and could cause the gene to become fixed in the population or even to spread.

As pointed out by the evolutionary biologists Robert Trivers and William Rice, this type of model is especially attractive for a gene on the X chromosome because an X-linked gene will be found twice as frequently in women, who have two X chromosomes, as in men, who have one X chromosome. As a result, the gene would have to increase childbearing in women only half as much as it reduced reproduction in men. Genes that hurt the reproduction of one sex while helping the other are not just a theoretical construct; they have been experimentally demonstrated in many different organisms and seem to play a key role in the evolutionary process of sexual selection.

How might a gene that makes men gay work to increase the reproduction of women? Trivers and Rice offer a simple explanation: If the gene did the same thing in both sexes, namely increased sexual interest in men, the women might be more likely to have intercourse and produce children. If this were true, then the sisters and maternally related female relatives of gay men should have more children than average, a hypothesis that might make for an interesting study.

SNEAKY MALES

Another way that "gay genes" could survive would be by increasing the reproduction of some of the men who carried it. This sort of model depends on the assumption that not every man who has the gene is actually gay, or at least not exclusively so. This seems to be a reasonable supposition since even gay brothers who shared Xq28 markers had fewer gay male relatives than expected for a simple Mendelian trait.

How could a "gay gene" help the reproduction of heterosexual male carriers? One possibility is suggested by the movie *Shampoo,* in which Warren Beatty plays a hairdresser obsessed with bedding his female clients. Their husbands don't suspect a thing because he is a hairdresser, so they assume he's gay. The silly premise actually makes some sense in a society in which dominant males protect their women, and the only males allowed in the harem are eunuchs. If men like the fictional hairdresser really existed, and "appeared" gay because they had the gay gene but were sexually attracted to women, they might help the gene survive.

A related possibility would be that the "gay gene" better enabled heterosexual carriers to survive to childbearing age, perhaps by helping them gather food or giving them better social skills to keep them out of trouble, thereby increasing the chance of the gene being passed on to future generations. This would be especially effective if a "side effect" of the gene was to make men more sexually active. If the gene worked like a sex trigger, everything would depend on the first shot. A young man who had his first sexual experience with a man would pull the gay trigger, but the young man first aroused by a woman would grow up to be heterosexual. The sexually active heterosexual would have many children and thus make up for the lack of offspring by the gay man.

Another way that a "gay gene" could survive would be through kin selection. If gay men helped their close relatives to produce or raise children, this might offset the smaller number of children that the gays themselves had. In order for the gene to prosper, a gay man would have to increase the reproductive success of his full siblings by more than twice what he lost himself. Although kin selection is an inefficient way to pass on genes, it's thought to play a role in the social behavior of some insects.

HYPERVARIABILITY

The final model of how a "gay gene" could survive evolution is that the gene actually has no selective advantage but stays in the gene pool because of an extraordinarily high rate of mutation. Such an "anti-Darwinian" theory would have been laughed out of the lecture halls a few years ago, but new discoveries about the human genome make it plausible.

Hypervariable DNA sequences change their structure at rates hundreds or thousands of times higher than the normal rate of spontaneous mutation. One of the best studied examples is the trinucleotide repeat sequence in the fragile X gene that is responsible for one form of mental retardation. In most people these sequences are present in 6 to 50 copies, but occasionally they elongate to 200 to more than 1,000 copies—an expansion that causes severe mental retardation and other abnormalities. The "full mutation" has survived and is present in about 0.1 percent of all X chromosomes, even though these people almost never have children, and several percent of all X chromosomes have a slightly elongated form of the repeats.

Evolutionary theorists have not yet come up with any good models to explain how such an obviously deleterious mutation can exist in the gene pool at such a high level. One guess is that there is some unknown advantage to having a flexible genome. The bottom line, however, is that DNA sequence variations that severely diminish reproduction can and do exist in the human genome. This model, unlike some of the other models, could be tested in a lab. If the "gay gene" is isolated and it is found to have a hypervariable DNA sequence, that could explain its survival.

SEXUAL EVOLUTION

Genes don't care about us. Some people think there is a "reason" for the "gay gene" or that it serves a larger purpose. On one side of the argument are those who believe homosexuality is nature's way of controlling the population. This can't be true, however, because the first people to die out would be the homosexuals, and with them the "gay gene." Another theory is that homosexuals somehow benefit society and therefore are "allowed" to survive evolution. The problem with this premise is illustrated by the joke about the professor who says, "If there weren't homosexuals, we wouldn't have opera." To which another professor replies, "Yes, but nobody would care." The point is that evolution works at the level of genes and individuals, not at the level of populations or societies. This may seem selfish—but then, genes always are.

Perhaps the most important question to ask is not "How could a 'gay gene' evolve?" but "Are there genes for sexuality in general?" Everything we know about evolution and natural selection says there must be. Since evolution selects genes that allow people to reproduce, there must be genes that encourage men and women to have sex and raise children. If that premise is accepted, and it is understood that the genome is constantly changing over evolutionary time, the question of how a "gay gene" could exist becomes one of precise mechanisms rather than of fundamental philosophy—a question of how the sexuality genes work, not whether they exist.

It is even more difficult to predict what will happen to a "gay gene" in the future than it is to trace its history. If gay men have fewer and fewer children through intercourse with women, the gene may die out. But if sex and reproduction are increasingly separated through the use of techniques such as artificial insemination, and if the "gay gene" has some beneficial "side effect," it may prosper. People might even *want* to introduce the gay gene into their offspring if the advantages were great enough.

Chapter Twelve

BEYOND SEX

The discovery of a genetic link to homosexuality is bound to be followed by discoveries of links to countless other aspects of personality. Many of the issues raised by our research, and many of the pitfalls, are the same, as are the techniques used to identify the genes. It's likely that finding a "gay gene" will be remembered as a breakthrough not so much for what it explained about sexuality as for opening another door to understanding genetic links to many equally complicated human behaviors and conditions.

Genetic research also will continue to play an ever greater role in understanding and fighting disease. Now a doctor seeing a patient for the first time routinely begins with a family medical history, the assumption being that the susceptibility to certain illnesses runs in families. In time, more of the actual genes involved in those diseases will be found and understood, and family histories will likely be supplemented by genetic tests.

Humans come in all shapes and sizes on the inside as well as the outside. Some think quickly and easily, others slowly and with difficulty. Some are happy, and some are sad. People who listen to their inner voices and dreams can end up being called creative, or being institutionalized. Some are good at math, others at poetry, and still others at deception. Some will lead healthy productive lives, while others will battle mental illness or debilitating addictions and obsessions. Cancer will strike some; heart disease will take others at a young age.

These different characteristics have two things in common: They are complex, meaning there is no single factor responsible for any of them, and they seem to be influenced to some degree by genes. Sexual orientation also is a complex trait, and this chapter will explore how the methods we used to study homosexuality might be applied to other complex human characteristics. Genetics won't provide all the answers for human behavior or cure every disease, but it may provide important clues to a wide range of psychological and medical characteristics.

AIDS AND KAPOSI'S SARCOMA

One of the main goals of our project was to identify genes that influence the progression of the AIDS virus, focusing on the cancer known as Kaposi's sarcoma. When we started, I assumed we would make quicker progress there than on sexual orientation because more background work had been done and likely genes already had been identified. Instead, the medical segment of the protocol turned out to be the toughest part because of problems finding the right families and because the genes we picked to test were not the best choices.

Other investigators had found subtle correlations between histocompatibility antigen genes (which influence the immune system) and HIV susceptibility and progression. Those studies were done on genetically unrelated people, however, so it was difficult to say if inheritance really caused the differences. Our plan was to look for a genetic link in pairs of brothers with the HIV virus, but it was hard to find brothers with exactly the same outcome from HIV infection, or even demonstrably different outcomes that could be contrasted.

Thirty-eight percent of the gay male subjects in our study whose HIV status was known were HIV positive, which is a higher rate of infection than for the rest of the gay population because many of the subjects were recruited from AIDS clinics. Among the gay brothers of the HIV-positive subjects, 19 percent were HIV positive and 81 percent were HIV negative. Although the HIV+/HIV− pairs could in principle have been used to detect genes for resistance to viral infection, in most cases the second brother probably was HIV negative not for any genetic reason but because he had avoided infection by minimizing risky sexual contacts. Only one pair had an HIV-positive subject with an HIV-negative brother known to have been exposed to the virus (via anal intercourse with men who later died of AIDS, including one who had Kaposi's sarcoma lesions at the time of sex). In early 1994, we were studying this pair for signs of genetic signals that say "stop" or "go" to the HIV virus.

There were five pairs in which both brothers were HIV positive, which was too small a sample for linkage analysis, especially for what is undoubtedly a very complex characteristic. We did try to see if two brothers were more likely to develop the same outcome of the virus. For example, if one brother got Kaposi's sarcoma, would his brother also get Kaposi's sarcoma, or would he get something else? If the brothers had similar outcomes, it might signal some genetic component. As far as we could tell from this limited sample, there was little obvious similarity in the progression of the virus in brothers. In one case, for example, both brothers had severely compromised immune systems, but one developed a rapidly metastasizing form of Kaposi's, while the other avoided cancer but was injected by the cytomegalovirus, one of the family of herpes viruses.

A complicating factor was that even if brothers shared the AIDS virus, they often received different medical treatments, which made it difficult to determine if outcomes were influenced by drugs, therapies, genetics, or a combination of these.

All of the families in which anyone was HIV positive were tested with more than thirty different genetic markers on several different chromosomes, but as might be expected from the small sample and the diversity of disease outcomes, no useful linkages were found. In early 1994, we began to seek better families for this type of genetic analysis and continued to track the health of our original subjects. Already during the course of the study, four of the HIV-positive men have died. I continue to believe that the body's own natural defenses may yet provide novel therapeutic strategies for combating AIDS, but finding these defenses will require research in many labs.

ALCOHOLISM

If we had trouble locating enough families to find a genetic link to AIDS progression, we located almost too many families with alcoholism and substance abuse. Of the gay male subjects who were interviewed about their drinking and substance use, 29 percent had a problem with drugs or alcohol. The rate was higher than normal in part because many of the subjects were recruited from treatment and self-help programs, but even when we didn't deliberately seek out such subjects, the rate was considerably elevated. For example, among the patients from the HIV outpatient clinic at the NIH, 47 percent had a history of alcohol or substance abuse. This may be more environmental than genetic, however, because many of the gay alcoholics and addicts blamed their behavior on feelings of rejection and alienation caused by their sexuality. Many also mentioned gay social life, which often is dominated by the bar scene and the use of drugs and alcohol to overcome feelings of anxiety.

Nor was there a problem finding relatives of our gay men with drinking or drug problems: Virtually every one of our addicted subjects named a blood relative with a problem. Determining patterns of inheritance was difficult, however, because people were more vague about the definition of alcoholism and addiction than of sexual orientation. One subject, for example, was sure a sister was a closet alcoholic because she drank two six-packs of Diet Pepsi every day. This means that any family study of alcoholism is unreliable unless all the relatives are personally interviewed, and even then the results are not as trustworthy as for sexual orientation.

Despite the difficulties measuring substance abuse, it clearly runs in families. For example, in one family the father was addicted to heroin, the mother was alcoholic, and every one of seven children had a documented problem with drugs or alcohol. Other studies have estimated that the son of an alcoholic has a ten-times-greater-than-normal chance of being alcoholic. Although many scientists blame this family clustering on genes, recent studies of twins suggest genes actually play less of a role in alcoholism than in sexual orientation. Our own family interviews showed numerous cases of how alcohol poisoned life at home for children and might have led them to drink heavily as adults.

The lack of good family information for alcoholism forced us to give up the idea of linkage analysis on relatives and try association analysis, which looks at the frequency of markers or mutations in unrelated people with the trait. We did have many individuals with an unambiguous diagnosis of alcoholism, and we thought we had a good place to look at their DNA: the dopamine D2 receptor gene, which had been correlated to alcoholism in some studies. A group of scientists in Texas had found an association between alcoholism and one particular marker near the D2 receptor gene, a finding replicated by at least two other labs. The evidence was not unanimous, however, and researchers at our own NIH and the Yale University School of Medicine found no relationship between the D2 receptor gene and alcoholism.

Even if the association were solid, it was unclear what it would mean. Some people argued that since the DNA sequence in question isn't even in the part of the gene that makes proteins, it is difficult to understand how it could influence alcoholism or anything else. Pablo Gejman, a fellow in Elliot Gershon's lab at the National Institute of Mental Health, decided to scan the entire gene for a mutation, and we gave him DNA samples from 13 gay men and lesbians with unambiguous diagnoses of alcoholism. After a year-long search, Pablo and his colleagues couldn't find a single mutation in the DNA from our 13 subjects. Nor did they find a significant mutation in another 161 alcoholics, including some who were in the original Texas study.

The story of the D2 receptor gene and alcoholism unfolded just the way the one about the androgen receptor and sexual orientation had: Both were receptors and reasonable candidate genes, but neither showed functional association. The lesson learned also was the same:

Not enough is known about the underlying mechanisms of these traits to permit anything more than guesses about what genes to explore.

We plan to continue testing our samples as new candidate genes are found, but meanwhile we did make one interesting finding about alcoholism. Researchers divide their subjects into Type I alcoholics, men and women whose social drinking becomes problematic in mid-life, and Type II alcoholics, antisocial men who start binge drinking at an early age. Some researchers point to the all-male Type II alcoholics and the correlation between drinking fathers and drinking sons as signs that a gene on the Y chromosome is involved.

In our study, however, all the gay male alcoholics fell into Type I. They didn't beat up anyone or smash cars and said they were more likely to drink quietly and alone. The surprise was that some of the lesbian subjects showed the characteristic signs of the more macho Type II alcoholism. One lesbian said that when she was 18 she was refused service at a bar, so she tore up the place on her motorcycle and drove back onto the street through the plate-glass window. Cases like this would seem to suggest that the gene cannot be located on the Y chromosome, which only men have.

Our most important epidemiological observation was the high rate of alcoholism in both gay men and lesbians, something that has been suspected but not systematically investigated. A careful study may be unpopular with conservatives, who don't like any "special" programs targeting gay men and lesbians, and equally so with some gay activists who worry about equating homosexuality with any disease or problem. Without more research, however, we won't know for sure what the connection is between sexual orientation and substance abuse, how serious it is, and what the solution might be.

GENES AND MENTAL ILLNESS

Most research on genes and human behavior (and funding for it) is directed at psychiatric diseases. In the past, mental illnesses were considered to be either a punishment from God or caused by an inability to cope with life's burdens. With Freud and the advent of psychotherapy, "madness" became an illness, usually blamed on emotional wounds to the psyche. The trend now is to view mental illness largely as a matter of biology, specifically, altered brain chemistry. The reason for the current view is the discovery that at least some mental diseases respond to drugs. Also, family, twin, and adoption studies have shown a substantial inherited component for many forms of psychopathology, including schizophrenia, major depressive disorder, panic disorder, Tourette's syndrome, hyperactivity, and anorexia nervosa.

The two mental disorders that have been studied most extensively at the molecular level are manic-depressive illness and Alzheimer's disease. The history of the first shows the wrong way to study complex mental traits, and the history of the second shows the right way. These case studies are important because they help explain why we chose the methods we used to look for genes related to sexual orientation, and they point the way toward how to study a genetic role in other behaviors.

Manic-Depressive Illness

This condition, also called bipolar disorder, is characterized by periods of deep depression alternating with periods of euphoria, grandiosity, and restlessness. The rates of manic-depressive illness in the identical twins of patients are about fiftyfold higher than in the population at large, and the rates in siblings are five- to tenfold higher than expected by chance, suggesting a substantial genetic component to the disease.

Unfortunately, early attempts to genetically map manic-depressive illness have been unsuccessful. One of the best known studies was carried out on the Older Order Amish of Pennsylvania. The Amish long had suspected manic-depressive illness could be inherited; the Pennsylvania Dutch phrase is *Siss im Blut,* "It's in the blood."

The Amish are a perfect laboratory for studying genes because they rarely marry outsiders and most of the genetic variability in the community can be traced back to the fifty or so founders who migrated from Europe in the early 1700s. They keep detailed genealogies, and their rigid conformity means that most grow up in a similar environment.

Geneticist Janice Egeland used those characteristics to identify one very large, extended family with numerous cases of manic-depressive illness. By 1984, when molecular mapping was still in its infancy, she had collected enough DNA samples to do a linkage study. She joined a group at MIT led by David Housman, one of the early pioneers in the field, and they analyzed the family, using the old-fashioned Mendelian approach, assuming that a single gene was responsible for the disease being handed down from generation to generation.

Much to their surprise, they almost immediately discovered a linkage to markers on chromosome 11. The LOD score was only 2, so Egeland went back and collected more DNA samples from the same family until the LOD score surpassed 3. The group then published the results to much fanfare in *Nature.*

Unfortunately, the finding soon fell apart. Two relatives who had been perfectly healthy at the time of interview developed unmistakable signs of bipolar illness. Since the analysis included both affected and unaffected relatives, the change caused the LOD score to plummet from apparently meaningful to negative. They also discovered that a separate branch of the family failed to follow the expected pattern. Further doubt was cast by other investigators who found no evidence for linkage with chromosome 11 in non-Amish people.

While Egeland was in Pennsylvania, another psychiatric geneticist, Miron Baron, was seeking a gene for manic-depressive illness in Israel among Jews of Middle Eastern descent. Following on the work of others, Baron found that twice as many woman as men had the disease and that fathers rarely passed it to their sons. These were exactly the characteristics expected for a dominant X-linked disease.

The best way to test the hypothesis would have been to map the DNA of families with manic-depressive illness using a series of X-chromosome markers, but at the time the genetic map was still rough, and Baron was impatient. Instead, he looked for families with both manic-depressive illness and the obvious X-chromosome traits of color blindness and glucose-6-phosphatase deficiency, both of which happen to be located at Xq28. Although these were the only markers tested, Baron tried to claim a tight linkage of these traits in four of his families and asserted an incredibly high LOD score of greater than 9, which meant odds of one billion to one.

This finding also was announced with great fanfare in *Nature,* and again other investigators failed to replicate the results. Baron tried to argue that maybe the results only held for the ethnic group he studied, but when he checked his own families with real DNA markers the results collapsed.

The solution to these research problems is not, as has been suggested, raising the statistical criterion to some magic number to guarantee accuracy, or adding new diagnoses. The answer is to do the analysis only on people who definitely have the disease and to use multiple, closely spaced genetic markers. To be safe, studies of psychiatric illnesses should, like our study of sexual orientation, concentrate on many small families instead of a few large families.

Alzheimer's Disease

There are examples of mental disorders that have been properly mapped. One of them, Huntington's chorea, is an incurable disease that slowly destroys the brain and is caused by a single, dominant gene on chromosome 4. Mental retardation is another problem for which several genes have been unambiguously identified. A more complex and scientifically interesting case is Alzheimer's disease, which is characterized by a progressive mental deterioration that begins with mild forgetfulness and ends in dementia.

People used to confuse Alzheimer's with "senility" and regarded it as something that happened randomly to some older people and not to others. Now research has shown it is a disease caused by the formation of distinctive plaques and neurofibrillary tangles in regions of the brain involved in thought and memory.

Alzheimer's disease is not caused by a single gene—there are many examples of identical twins who don't share the disease—but research has found a substantial genetic component involving at least three genes. The first gene is involved in the rare cases where the disease strikes during middle age. Analysis of the brain plaques in these sufferers implicated a protein called beta-amyloid, which is produced by a gene on chromosome 21, the same chromosome involved in Down syndrome. In several families, all the younger victims of Alzheimer's had a mutation in the middle of the beta-amyloid gene.

Other families showed no linkage to chromosome 21, however, and the search continued until a second locus was identified on chromosome 14. So far the actual gene there has not been isolated.

Another researcher, Margaret Pericak-Vance at Duke University, set out in 1991 with a different approach. She looked for genetic linkages only in people with Alzheimer's in smaller families, and she found a weak but significant linkage on chromosome 19. If she had included people without the disease in the analysis, as every scientist before her had done, the linkage would not have appeared so significant. As it turned out, her finding proved to be the most important to date for the genetics of the disease. Unfortunately, people in the field paid little attention because they had their minds locked onto big families with more obvious Mendelian patterns of inheritance. Besides, they said, chromosome 19 couldn't have any of the "right" genes for Alzheimer's because there was nothing known there that related to the brain.

Another finding suddenly changed the conventional wisdom. Scientists discovered that the brain-clogging amyloid plaques contained a molecule called apolipoprotein E, or apoE for short, a protein that carries cholesterol in the blood. The surprise was that the gene for apoE was sitting on chromosome 19, exactly where Pericak-Vance had found linkage. New studies were done, and it was discovered that people with two copies of one version of the gene, dubbed apoE4, had a 91 percent chance of developing Alzheimer's disease if they survived to age 85, while people without the E4 version had only a 20 percent chance of developing the disease.

There are two fascinating lessons from the Alzheimer's story. First, it proves that seeking out relatively rare large families that show Mendelian patterns of inheritance will turn up rare genes (the linkages on chromosomes 21 and 14), but the better strategy for finding more common genes is to look at small families, using the shared-trait method. The second lesson is that no matter how much has been written about a trait or a disease, there is no way to predict what sort of genes might be involved. No one ever would have guessed that senile dementia could be caused by a fat-carrying molecule in the blood. For our study on sexual orientation, the lesson is that a "gay gene" might perform some role we have never imagined.

GENES AND WHO WE ARE

To understand how genes help shape who we are, rather than just what we suffer from, it helps to start with something simple and move to more complex areas of behavior. The first example is something most people take for granted.

Handedness

Most people would assert they were "born" right-handed or left-handed. Indeed, infants favor one hand over the other at an early age, and attempts to change the preference are difficult and even harmful. Furthermore, it doesn't appear to be something learned, because children who are adopted by left-handed parents don't have an increased chance of being left-handed. By contrast, the blood child of a left-hander does have an increased chance of being left-handed, and the child of two southpaws has an even greater chance.

Surprisingly, when the heritability of handedness was measured, it was found to be only about 30 percent, which actually is less than was found for sexual orientation. The calculation comes from the fact that most of the identical twins of left-handers are right-handed, and that other siblings of left-handers are only twice as likely as the rest of the population to be left-handed.

There are various theories about why being left-handed is not more easily inherited, such as blaming it on early brain injury, but an alternative explanation is that there is a single "handedness gene" that only affects some people. According to this model, advocated by the English scientist I. Chris McManus, individuals with two copies of the "sinistral gene" have a 50 percent chance of being left-handed, people with one copy of the "sinistral gene" and one copy of the "dextral gene" have a 25 percent chance of being left-handed, and people with two copies of the "dextral gene" are almost always right-handed. This model remains to be tested, but it shouldn't be hard to find several hundred pairs of left-handed siblings. Finding such a gene could offer new insights into one of the most fundamental and uniquely human aspects of how our brains are organized.

Personality

Finding a gene for handedness would be easier than finding a gene for more complex parts of our personalities, but that doesn't mean those genes don't exist. Some of the most widely known evidence for a genetic role in personality comes from studies in Minnesota, in which twins who were separated at birth are located and brought together as adults. Some of the results seem too absurd to be true: men who haven't seen each other since birth who followed the same careers, married women with the same first names, and drove the same model cars; girls who both sprained ankles at 15, who were afraid of heights, and who drank their coffee black, no sugar, and cold. The evidence supporting the role of genes in personality is more than just anecdotal, however, and it is supported by a large number of careful, systematic studies examining twins, siblings, parent-offspring pairs, and adoptees.

Webster's defines personality as the "sum total of the physical, mental, emotional and social characteristics of an individual." One way that personality can be studied in groups of people is to separate the individuals by the "superfactors" of neuroticism and extroversion.

Neuroticism is a general measure of emotional instability or maladjustment, rather than a particular neurosis. People who score high on the neuroticism factor tend to be anxious, moody, hostile, and depressed. They are unable to cope with stress and may panic or feel hopeless when faced with an emergency. More than 25,000 twin pairs have been studied for the neuroticism factor, and a meta-analysis of this data indicates that heritability is about 30 percent. Adoption studies suggest that in these cases most of the similarity between relatives is due to genes rather than specific family environments. For example, the correlation between a biological parent and child is about 15 percent compared to less than 1 percent for an adoptive parent and child.

The second superfactor, extroversion, is a measure of sociability. Extroverts tend to be cheery, assertive, active, and talkative. They like large groups and prefer excitement and stimulation to peace and quiet. Twin and family studies of extroversion have found results very similar to those for neuroticism: overall heritability is about 30 percent, and most of the familial resemblance can be traced to inheritance.

The fact that both neuroticism and extroversion are substantially genetic does not mean there is a single "neurosis gene" or "extroversion gene." In fact, the available data suggest that both facets of personality are probably influenced by multiple interacting genes. The evidence is that identical twin correlations for these traits are typically five to ten times higher than the correlations for fraternal twins, siblings, or parent-offspring pairs. In this case, the expression of the personality trait requires not just one gene but many genes acting together. This will make looking for the personality genes by linkage techniques difficult but not impossible.

Several more specific dimensions of personality have been studied by comparing twins and comparing biological with adopted children. The results show that the traits of "absorption," or imagination; "aggression"; "alienation"; and "social potency," or leadership, all display correlations of greater than 50 percent in identical twins regardless of whether the twins are raised apart or together. The implication is that many aspects of human character are deeply embedded in the genetic makeup.

Faint traces of genes can be found even in traits that appear to be entirely cultural. For example, several studies have found a substantial genetic component for "traditionalism." The role of genes in this trait may be exacerbated by assortative mating, which is the tendency for "birds of a feather to flock together"; i.e., of conservatives to marry conservatives and of liberals to marry (or at least have children with) liberals. The possibly genetic tendency to follow rules and authority, however, should not be confused with what particular rules or authorities are followed. For example, similar studies have shown that there is no genetic influence on belief in God or attitudes toward racial integration.

Not all personality traits have a genetic component. Intimacy, for example, shows very little heritability. In terms of the sexual orientation story, one of the most interesting findings is that traditional measures of masculinity and femininity show just about the same correlation in identical as in fraternal twins, a sign of low genetic influence. That seems to support our finding that the "gay gene" is not a "sissy gene."

Shyness and Aggression

Some of the most intriguing evidence for the role of genes in human personality comes from studies of opposite traits: shyness and aggression.

As nursery school teachers well know, children differ markedly on their first day of school. Most youngsters are upset to be separated from their mothers at first but soon adjust to the new children and toys and begin playing. Others, however, seem inconsolable at being separated from the parent, and even after being pried away are uneasy and afraid of the new sights and sounds.

Jerome Kagan, a developmental psychologist at Harvard University, has found that these differences appear well before the first day of nursery school. He believes that about 20 percent of children are born "temperamentally shy" and can be picked out by a number of physiological signs that appear during the first year of life. These include an increased heart rate while asleep, a distinct pattern of brain waves, increased blood pressure in response to new situations or stimuli, and an unusual sensitivity to allergies and hay fever. Kagan has hypothesized that these are due to altered brain chemistry in the amygdala, an almond-size structure that regulates the sympathetic nervous system and generates the racing heartbeat, perspiration, and dry mouth that are signs of fear. As the children grow, their physical differences translate into the type of behavior we call shy or timid. They are quiet and reserved with other children, dislike noisy or competitive play, and try to avoid novel places and things.

There is not any single "shyness gene" that explains why some children are withdrawn and others outgoing. Nevertheless, family and twin studies suggest a substantial genetic component. Robert Plomin, one of the leading theoretical and experimental behavioral geneticists, has noted that "Studies suggest that shyness is the most heritable component of personality," a claim based on heritability estimates of about 50 percent.

Not all temperamentally shy children will be shy adults. In fact, Kagan estimates that less than half of the youngsters he has studied remain bashful and withdrawn in adulthood. Conversely, many children are shy not because of some underlying biological reason but because of disagreeable or frightening experiences early in life. Notwithstanding these caveats, the connection between one form of shyness, biological markers, and genes seems robust and might be worth studying at the molecular level.

Just as the nursery yard will have its shy children, so it will have its bullies. And as the traditional schoolyard bully is replaced by today's gun-toting gangster, the question of the origins of human aggression and violence are of increasing social as well as scientific importance.

Hans Brunner, a Dutch scientist, has shown that one very rare form of aggression in men is genetically determined. Brunner and colleagues studied a large family whose men had a long history of violent outbursts. One of them was a rapist, two were arsonists, and one tried to run over his boss with a car. Their family tree shows the classic hallmarks of recessive X-linked inheritance: The trait is found only in men and is passed on through their mothers.

This pattern led the scientists to perform DNA-linkage analysis with a series of X-chromosome markers and gene probes. They found that all the aggressive men in the family had a single mutation in the gene for the enzyme monoamine oxidase in the middle of the long arm of the X chromosome. The mothers of the aggressive men had one copy of the mutant gene and one copy of the normal gene, but the nonaggressive men had only the normal gene. That pointed to a simple, one-to-one correspondence between the mutation and aggressive behavior —a situation similar to the relationship between the X-linked color pigment genes and color blindness and unlike the more complex correlation between Xq28 and sexual orientation.

Although the cause-and-effect relationship between the monoamine oxidase gene mutation and aggressive behavior seems clear, it is important to realize that this particular mutation has so far been identified in only a single family in the entire world.

Other research has suggested that people with low levels of the chemical serotonin are prone to both impulsive, violent acts and to depression and suicide. (Although this might seem contradictory, psychologists have long postulated that aggression and depression are linked; one is anger directed without, the other anger directed within.)

Low serotonin by itself is probably not hazardous, but when combined with stress, alcohol, or abnormal levels of noradrenaline, it might lead to impulsive reactions as extreme as murder.

Markku Linnoila, scientific director of the National Institute on Alcohol Abuse and Alcoholism, has been looking for genetic links to low serotonin levels and violence. Early studies, which found a correlation between violent behavior in fathers and sons in Finland, suggested an inherited component. More recently, Linnoila has used molecular techniques to find a possible genetic marker in Scandinavian men who display antisocial behavior, alcohol-induced violence, and multiple suicide attempts.

The marker lies within the gene for tryptophan hydroxylase, an enzyme that converts the amino acid tryptophan into serotonin, and is therefore a reasonable candidate gene for aggression. It should be emphasized, however, that so far no mutation within the enzyme coding sequences has been found, and that the findings remain to be replicated.

These types of studies suggest that genes can play a role in both shyness and aggression, at least in some people under certain circumstances. The findings should not be taken to diminish the role of life experiences, upbringing, economic status, or class, however. Rather, they widen our appreciation for the diverse components that contribute to how we get along with our fellow human beings.

IQ

A dictionary definition of intelligence is "ability to adapt"—a characteristic that is difficult if not impossible to measure in a laboratory or classroom. Instead, psychologists attempt to measure cognitive abilities with paper-and-pencil tests that yield an overall intelligence quotient, or IQ score, which compares the performance of an individual to others the same age.

Since the 1920s, hundreds of family, twin, and adoptee studies of IQ have been reported in the scientific literature. The main conclusion is that roughly half of the variation in IQ scores is caused by genetic differences. This figure is determined by comparing correlations between individuals with different degrees of genetic relatedness and with the same or different rearing environments. For example, the average correlations between genetically identical twins is 86 percent if raised together and 72 percent if raised apart. For siblings, who share half of their genes, average correlations are 47 percent if raised together and 24 percent if raised apart.

The study of adoptees produces a separate estimate of environmental influences. For example, the correlation between genetically unrelated children who are reared together is 34 percent. As might be anticipated, the size of the environmental component is largest for twins and smallest for cousins.

Together these findings show that both heredity and environment contribute to general cognitive abilities—a conclusion that reinforces the idea that both nature and nurture can contribute to the same trait. Although there is a clear genetic influence, that doesn't mean it will be easy to find the genes. Except in extreme cases, such as mental retardation, IQ is very much a continuously distributed trait, meaning it comes in an infinite range of shades rather than a few distinct colors.

Moreover, the twin and family studies suggest that identical twin correlations are more than twice sibling correlations, a finding that suggests multiple interacting genes. Thus current attempts to find genes that are associated with IQ scores focus on individuals at the extremes of the spectrum. The hope is to identify locations where genes contribute in a quantitative manner to cognitive ability—not to find some single gene that flicks on and off between smart and stupid.

GENES AND BEHAVIOR

Many people are disturbed by the idea that genes influence behavior, and books have been written to try to discredit the theory. Public distrust of behavioral genetics research and inflammatory reporting in the news media have had a chilling effect on research and progress in all areas of behavior genetics, not just sexuality. Most of the objections are based on a few misconceptions about what this research is attempting to do.

"Genes are destiny." This common misconception is most avidly professed by people who don't understand how genes work. They visualize genes as "master puppeteers" rather than as what they really are—chemical structures that code for protein production or regulate the activity of other genes. As noted by Robert Plomin and his colleagues in the textbook *Behavioral Genetics,* there is no such thing as a gene "for" a particular behavior, any more than there is a gene "for" the length of one's nose. Rather, genes influence behavior through indirect and complex paths that require inputs from physiology, the environment, society, and culture.

The reason that humans are so marvelously varied and adaptable is that we do not arrive ready made; some assembly is required. The genome may contain the directions for the assembly of the brain, but we have great influence over how the pieces are put together. For example, even a person who has a gene "for" alcoholism can live as a teetotaler.

A related mistake is the idea that genes exert a constant, equal influence on behavior, from conception till death. Just as the genes responsible for physical development are active only during brief, specific periods, genes that influence behavior may have strikingly different effects over the course of a lifetime. For example, a twenty-year study of IQ in twins showed that genetic effects accounted for only 15 percent of the variance in mental abilities in infants but 40 percent of the variance by elementary school age; other studies suggest that the heritability of IQ continues to increase throughout adulthood into old age. This area of inquiry is so promising it has spawned a new discipline called developmental behavior genetics.

"Genes make us all alike." When people hear a trait is influenced by a gene, they sometimes assume it only comes in two opposite varieties: smart or stupid, mean or kind, passive or aggressive. We know that physical characteristics such as height, weight, and skin color don't come in just two varieties, so we must assume that the variations of personality are equally vast. Just look in the mirror. Facial features are largely determined by genes, but there is nobody else in the world (unless you have an identical twin) who looks quite like you. Given the wonderful diversity with which genes sculpt the human face, they must be equally dexterous and imaginative when it comes to shaping the nooks and crannies of the human brain. Such breathtaking detail on the outside of the human form suggests an equal, or even more elaborate, construction on the inside.

"My ethnic group is better than yours." Just because genes contribute to differences between individuals does not automatically mean they contribute to differences between groups. The contribution of genes to individual variations in IQ is one of the most thoroughly researched and best documented findings in human behavioral genetics, but it would be wrong to jump to the conclusion that any differences between groups—such as ethnic and racial groups—are the result of genes. It is equally possible that differences between groups are caused by differences in environment—which in the case of IQ scores account for fully half of the differences between individuals. The same is true for other behaviors, such as violence. If people in Indiana commit more crimes than people in Illinois, for example, nobody worries about "Hoosier genes" weakening the stock.

"Genetics will lead to eugenics." Understanding the role of genes in human behavior does not have to lead to the deliberate manipulation of the human gene pool through selective breeding, abortion, sterilization, or genetic engineering. Attempts to interfere genetically with traits that are not life threatening are in my opinion both unwise and potentially dangerous. Genetic manipulations are hazardous to individuals because the results are too unpredictable; they are hazardous to the evolution of the species because, in our shortsightedness, we may eliminate genes that have long-range benefits; and they are hazardous to our own vision of humanity in all its glorious and beautiful diversity.

Humans have a strong urge to predict and manipulate the future. Will the astrologers and seers of the past be replaced by white-coated scientists and genetic counselors? Will a small group of affluent First-Worlders create a genetocracy, while the bulk of humanity is left to reproduce the detritus of unwanted DNA? The possibilities seem far-fetched, but the forced sterilizations encouraged by the early-twentieth-century eugenics movement make it clear they are not impossible.

It is equally important to realize, however, that throughout history it has been prejudice rather than technology that has been the driving force behind attempts to "improve" the human race. The scientific study of genetics did not cause the mass extermination of European Jewry during the Holocaust; the cause was anti-Semitism. If Mendel had never lived, and if genes had never been discovered, it is doubtful that a single additional Jewish life would have been spared.

A more recent threat is the abortion of female fetuses, solely because of their gender, which is well documented in developing countries, such as India. Critics blame prenatal testing, arguing that we'd be better off if ultrasound tests had never been invented. This ignores the great benefits of appropriate prenatal testing, but the main point is that it's not the test that causes the abortions, it's prejudice against women. Before the tests were available, people used another technique: They killed the girls soon after they were born. The enemy here is prejudice, not science.

"Genes are un-American." This idea confuses difference with inequality. The American idea that everyone is created equal does not mean that everyone is created the same or has the same genes. Our whole concept of equality is based upon respect for individuals despite their differences. If everyone were the same, we would have little need for the laws and precepts upon which democratic societies are built. Genes do contribute to individual differences in behavior, just as they do to variations in physical attributes. The question is how we, as a society, interpret and act on these differences.

Chapter Thirteen

BEYOND THE LAB: IMPLICATIONS OF A "GAY GENE"

The CNN reporter moved an inch closer, put a serious look on his face, and peered deeply into my eyes. I knew what question was coming next, the same one that every reporter asked.

"Finally, Dr. Hamer," he said, lowering his voice to emphasize the weightiness of the matter, "What are the *implications* of your research?"

What he meant, of course, were not the scientific implications, of which there were many, but the impact on society and politics, the legal system and medical ethics, and on the lives of real people.

I looked back at him and told him the truth: "I don't know."

To answer his question accurately I'd have to be more than just a scientist. I'd have to be a sociologist, politician, lawyer, and ethicist—not to mention a prophet, seer, and fortune-teller. It's not that I don't think about these issues and worry about the outcomes, just that I plain don't know the answers. Nor does anyone else. What I do know is that the reaction to our finding was strong—positive, negative, and sometimes hysterical—and that it spanned the disciplinary spectrum from politics to medicine, sociology, religion, and philosophy.

THE PRESS REACTION

The mainstream response, especially in the news media, generally was careful and responsible. Before the study appeared, I had worked with the public relations office of the NIH to help them prepare for the questions I knew would come. This also helped the reporters, who usually were on deadline, and except for the ones writing for the larger papers, didn't have science backgrounds.

I worried about stories heralding the discovery of a "gay gene" that automatically determines who is gay and who is straight. In most cases that didn't happen. Most reports were careful to explain that the linkage to Xq28 was just part of the larger story of sexual orientation. A very few news reports misinterpreted the finding to make homosexuality seem like a disease that could be "cured," but many other stories included quotes, such as the one from George Neighbors, Jr., the spokesman for the Federation of Parents and Friends of Lesbians and Gays (PFLAG), who said, "If you believe in God or Nature, that's what homosexuality develops from."

William F. Allman, in the 26 July 1993 issue of *U.S. News & World Report,* wrote about the study in broader terms: "Ever since Copernicus removed Earth from the center of the Universe, society has continually had to redefine what it means to be a human being . . . Self-knowledge is indeed unsettling, even dangerous, but the continued gathering of such knowledge is fundamental to the existence of the human species."

As was to be expected, the television stories were far more interested in the social implications than the scientific ones. I politely declined an invitation to appear on "Nightline," until they explained they were going to interpret the science with me or without me. I agreed to appear on the condition that I only would talk about the science, not the politics.

That was not to be, of course, and sure enough I soon heard Ted Koppel saying, "We are in the middle of a visceral and very tough political debate on the issue of gays in the military. The president is supposed to be coming out with a decision sometime in the next few days. If it hasn't been raised already, someone is going to raise a question about the timing of the release of your study. Did it have anything at all to do with this ongoing debate?"

"Well . . ." I began, thinking about two years of interviewing subjects, drawing blood, testing markers, analyzing the results, and struggling through the review process, . . . "no.

"We submitted our paper to *Science* when we had reached an appropriate statistical confidence level," I added dryly. What I really wanted to say was "Ted, that's the stupidest question I've ever heard."

The second guest on the show was Rev. Peter Gomes, a professor of Christian morals at Harvard University. Although he is a religious scholar, he was more positive about our work than his colleagues in the biology department had been.

"Part of the whole conversation about homosexuality has been to confuse it with some deliberate choice of lifestyle and to suggest that it is somehow an option which other people who are 'normal,' as it were, do not have," Gomes said. "In the sense that homosexuality is now to be seen, at least in part, as part of the equipment with which some people are born into the world, I think that's helpful."

POLITICIANS: LEFT AND RIGHT

Most conservatives were not happy with the study, mostly because they believe that being gay is a "lifestyle" that a person chooses, not something influenced by genes. The problem that some in the Christian right have in accepting a genetic role is that genes come from God, and they don't believe God wants people to be gay.

Louis Sheldon, a Presbyterian minister who appeared with me on the "MacNeil/Lehrer News Hour," had his own explanation. "Homosexuality is not genetic," he said. "Basically it is a very simple thing, so simple it's complicated. The body parts don't fit. I don't believe Nature would do that to the human race."

A conservative U.S. senator, in a letter to the NIH that his office released to the press, was upset that the National Cancer Institute was "diverting resources from cancer research to sexual orientation projects."

The NIH responded that the sexual orientation study was part of a larger protocol that included research on a specific cancer and other areas of interest to the institute, a protocol that had been properly and thoroughly reviewed. The results for the medical segment of the protocol had not been completed, the NIH said.

Some of the criticisms were more personal and underhanded. One antigay "psychologist" talked his way into our building by claiming to be a reporter. He asked Stella Hu about her personal life and started quizzing other people about their sexual orientation, until Vicki Magnuson caught on and threw him out of the lab.

Some progay activists also were upset with our study. Probably it's a good sign of balance, but some complained that looking for a "gay gene" would mean that homosexuality was being equated with a genetic defect. More apocalyptic warnings were that prenatal screening would lead to "genocide" against future generations of gays. Other gay leaders pointed out that recognition of a genetic basis for race hasn't helped African-Americans or other minorities overcome prejudice.

A psychology professor from California, for example, was quoted in the gay press as saying the study was irrelevant. "This is a political question, a moral question. The real issue is what are you apologizing for? You either think it's wrong or not."

Gregory King, spokesman for the Human Rights Campaign Fund, the country's largest gay rights organization, disagreed: "We find the study very relevant, and what's most relevant is that it's one more piece of evidence that sexual orientation is not chosen."

THE COURTS

The findings not only made it onto the talk shows, they almost immediately were introduced as evidence in the courts. I was subpoenaed to testify in Denver in the legal challenge to Amendment 2, an initiative passed by the voters of Colorado on 3 November 1992, prohibiting the state from giving homosexuals protected status or claims of discrimination before the law.

Opponents of the measure called it unconstitutional and said it would give a green light to discrimination against gays. Supporters said gays should not get "special" treatment, because they are not like racial minorities or other groups currently protected by law.

The opponents reasoned that if they could prove being homosexual is a deeply ingrained characteristic—and better yet, genetic—the courts would have to protect gays' civil liberties and access to the political system. They figured that a scientific finding linking homosexuality to genes was a perfect piece of evidence. When the antiamendment lawyers called for more information, I explained that our study did not show homosexuality was a *purely* genetic characteristic like race, that the findings only applied to men, and that our results hadn't been replicated. I was subpoenaed anyway.

So on 14 October 1993, I found myself on the witness stand in Courtroom 19 of the District Court of the City and County of Denver, State of Colorado, Judge H. Jeffrey Bayless presiding.

During the first hour of direct examination by Jeanne Winer, an attorney for the plaintiffs challenging the amendment, I explained that our study found a greater than 99 percent probability that sexual orientation was genetically influenced in at least some men. I also outlined how earlier twin and brain studies were fully consistent with our molecular results and said that there was no scientific evidence that sexual orientation was a trivially alterable trait.

The cross-examination was led by Assistant District Attorney Jack Wesoky, who wanted to show that gays as a group did not deserve to be protected by law. Mr. Wesoky had really boned up on his biology, and he jumped bravely into the technical issues of molecular genetics.

Reading from careful notes, Wesoky expressed surprise when I told him that the difference between the DNA of one person and the next was about 0.1 percent. He seemed incredulous that such a "small" amount of genetic difference could explain so many differences between people, including their sexual orientation.

I told him that 0.1 percent might not sound like a lot, but it actually corresponds to three million chemical base pairs, which is a very large difference.

Apparently sensing a weak point, he asked if that mere 0.1 percent difference could account for the fact that he is short and I'm medium height. I told him yes. Then he asked whether that little 0.1 percent difference could account for the fact that he is bald and I'm not; I told him that was part of the reason, but age was also important. Still pressing, he asked whether that 0.1 percent could account for the fact that he has a bigger bone structure than me, and I explained that three million genetic differences were more than enough to explain that minor discrepancy in our physiques.

What worked best was when I told Wesoky that although the difference between us was "only" 0.1 percent, the difference between him and a chimpanzee was not much larger, about 1 percent. That got a good laugh from the spectators.

What really brought down the house, though, was when Wesoky was grilling Professor Richard Green of the UCLA Medical School about Simon LeVay's discovery of a difference between gay men and straight men in a part of the brain.

What Wesoky wanted to know was whether the size of the hypothalamic nucleus observed by Dr. LeVay could have been influenced by the "male sex hormone, testosteroni."

That's testosterone with a hard *e*—or as one of the lawyers joked later, "You know, 'Testosteroni, the San Francisco Treat.' "

On 14 December 1993, Judge Bayless found Amendment 2 to be unconstitutional. Concerning the issues that I was asked to testify about, the judge wrote: "The preponderance of credible evidence suggests that there is a biologic or genetic 'component' of sexual orientation . . ." He declined, however, to decide the issue of "nature vs. nurture," an issue he said didn't matter in this case and was best left to another forum.

Gay-rights lawyers say they will continue to press the immutability argument in cases of gay marriage, sodomy laws, and military policy, not because they necessarily want to, but because it has become an integral part of civil-rights law. As long as opponents of gay rights continue to argue that homosexuality is a "choice," gay activists likely will continue to quote scientific findings of a biological role in sexual orientation.

BIOLOGY AND US

More than once, sitting in that Colorado courtroom, I thought, What am I doing here? What does biology have to do with law, politics, and legal rights, anyway? When it comes to figuring out how the world works, the scientific method is a formidable tool, but when it comes to deciding how the world *ought* to be run, there's a lot to be said for the rule of law, common decency, and morality.

Concepts like "good and bad" or "right and wrong" do not appear in biology textbooks. Nevertheless, there seems to be an almost irresistible urge for people to try to use biology either to condemn or justify homosexuality (and other human behaviors). They talk about what's "natural" and what's not, or what makes evolutionary "sense." None of the arguments really resolves the issue, however.

I received a wonderfully written letter from a man who had grown up in the country. "Animals don't do it, so why should we?" he asked. He worried that "You guys buried in your labs, especially ones in urban areas, function under the disadvantage of being divorced from nature." He asked, "Can you even imagine a queer grizzly bear? Or a lesbian owl or salmon?" In conclusion, he said that "If none of the lower orders engage in sex with the same gender, the motivating factor for homosexuality in humans must not be genetic, rather it must be in the noggin." Some gay supporters use the reverse of this argument, claiming evidence of homosexuality in sea gulls and sheep, for example.

Personally, I don't see why people are so interested in what happens in the barnyard; sometimes it seems they are more interested in how animals have sex than how we do. The fact of the matter is that there is no good animal model of human *heterosexuality,* let alone homosexuality. Pigs don't date, ducks don't frequent stripper bars, and horses don't get married.

Anyway, since when are animals good role models? The praying mantis devours her mate while they are still copulating. Male dogs will mount their daughters. Animals don't speak, write love songs, build churches, or do a lot of other things that we consider most worthwhile.

Another argument states, "Alcoholism might also be linked to genes, but that doesn't mean we should encourage people to drink." Unlike homosexuality, alcoholism is considered a disease because it harms a person's physical and mental well-being. Although some people regard homosexuality as a "disease" because some gay men die from AIDS, what's unhealthy is unsafe sex of any variety, not sexual orientation. A similar argument is "Suppose pedophilia is genetic. Does that mean we'll give 'special rights' to child molesters?" The answer is that pedophilia is wrong because it hurts an innocent victim. Even if pedophilia were genetic (and there's no evidence of this), it still would be a crime, just as violent aggression and murder are crimes, regardless of the motivation.

In short, biology is amoral; it offers no help distinguishing between right and wrong. Only people, guided by their values and beliefs, can decide what is moral and what is not. My own guiding principle is simple: Everyone has a right to life, liberty, and the pursuit of happiness, so long as they do not infringe upon the rights of others. From my point of view, this and this alone is the proper perspective from which to judge the morality of sexual orientation or any other aspect of human behavior.

PREVENTING MISUSE OF GENETIC RESEARCH

Biology and genetics may not provide the principles to resolve moral issues, but they do raise moral issues of their own, especially as the technology threatens to advance beyond our ability to understand all the implications. In terms of our research on sexual orientation, the most pressing questions are those concerning the development of genetic "tests" or "therapies" for homosexuality. I insisted—in the face of some opposition—on ending our *Science* paper with the following paragraph:

> Our work represents an early application of molecular linkage methods to a normal variation in human behavior. As the Human Genome Project proceeds, it is likely that many such correlations will be discovered. We believe that it would be fundamentally unethical to use such information to try to assess or alter a person's current or future sexual orientation, either heterosexual or homosexual, or other normal attributes of human behavior. Rather, scientists, educators, policy makers, and the public should work together to ensure that such research is used to benefit all members of society.

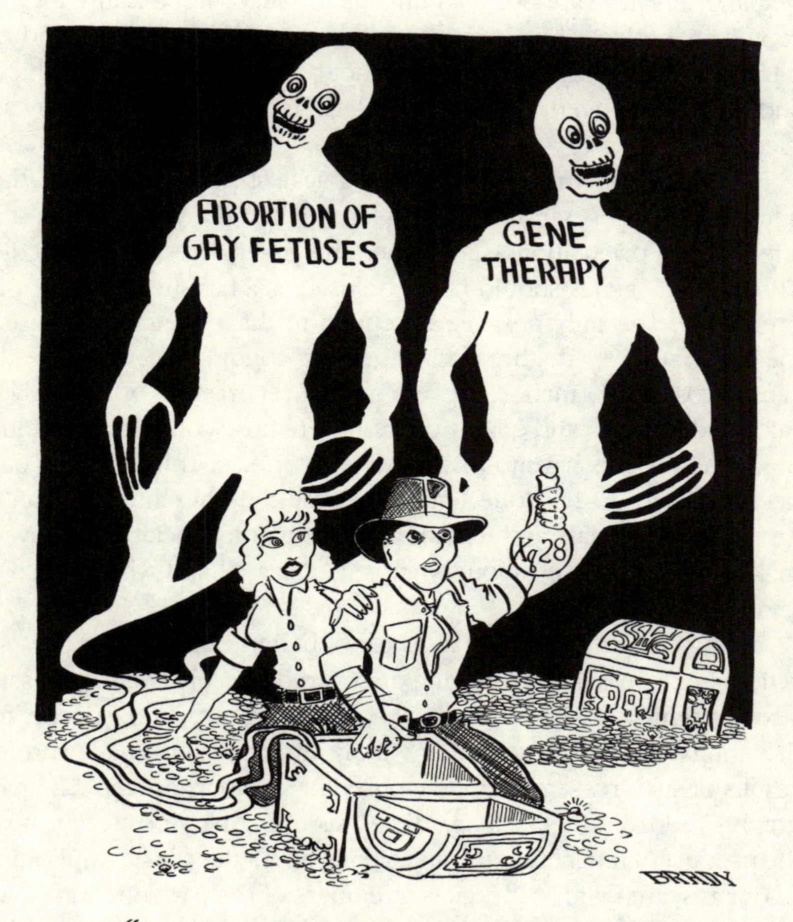

"AT LAST, DR. HAMER... THE SECRET OF THE GAY GENE!"

David Brady

There are three reasons I'm opposed to the genetic manipulation of human sexuality. First, I believe that both heterosexual and homosexual orientation are normal human attributes—not because they are rooted in biology, but because they pass my "life, liberty, and the pursuit of happiness test." That may not be very scientific, but ethics is and always has been a matter of beliefs and values, not of facts and theories.

Second, I think that discriminating against people based on their genetic makeup is wrong. That's as true when the genes in question affect sexual orientation as it is when the genes determine skin color. The "wrong" genes should never be used as a basis for terminating a pregnancy. The only possible exception to this rule is in the case of the most severe, life-threatening genetic conditions, i.e., those that cause gross malformations, severe mental retardation, or a greatly diminished life span and quality of life. When those conditions are found in a fetus, parents have to make a difficult choice. Third, genetic engineering is not to be done lightly because the consequences can be great for individuals and the species as a whole. Society will have to make difficult choices about what is a "disease" and what should be treated.

Being ethical is not only a matter of beliefs but of taking personal action and responsibility. Having discovered a link between genes and sexual orientation, I am accountable for the results, and I deeply feel the obligation to do everything within my capacity to make sure the results of such research are not abused or misappropriated. The paragraph cited from the *Science* article was not a hasty postscript. It was the product of much thinking and talking about the ethical implications of our research with colleagues, members of the gay community, and my advisory committee—a group that included two bioethicists, a lawyer, and a minister. It also was meant to be my first act in this area, not my last.

The ethical issue raised most often about our research is that it might, hypothetically, lead to a blood test that would reveal a person's sexual orientation. Presumably the test would scan the DNA for all possible variations of genes involved in sexuality to create a "genetic profile" that could be compared with the profiles of people of known sexual orientation. It would be the genetic version of the personality tests given to see if job applicants will "fit in" to a given company.

Insurance companies might be interested in such a test if they decided that gay sex should be considered in actuarial tables along with driving records, cigarette smoking, or hobbies like skydiving. The test might also be of interest to the military, which has had trouble defending its distinction of allowing gay people but not gay conduct.

The second possibility raised by our research would be a similar test used with amniocentesis or any kind of prenatal exam to determine the genetic makeup of a fetus. Presumably this test would be taken by mothers who could then elect to abort a fetus if they thought it might grow up to be gay.

The third and most remote possibility would be some sort of genetic therapy whereby the hypothetical gay gene or genes were replaced by their heterosexual counterparts. Theoretically, this could be done in the womb or even applied to the egg or sperm before fertilization takes place.

Imagine the case of Suzanne, who is pregnant with her first child. At the urging of her husband, she takes a new experimental prenatal test that shows her fetus has a 90 percent chance of being gay—just like her brother. This causes a great deal of anguish and soul searching, and in the end, Suzanne decides on an abortion so she can try again for a heterosexual child. The abortion is botched, however, and she will never be able to have another child.

That story actually is the plot of a play called *The Twilight of the Golds,* which, ironically enough, opened at the Kennedy Center in Washington, D.C., just a few weeks before our paper appeared. Playwright Jonathan Tolins was called a visionary, or at least a terrific marketer, but so far there is no real-life prenatal test for sexual orientation.

In fact, the results that we published do not allow scientists to learn anything about the sexual orientation of an individual, either living or unborn. This is because we did not actually isolate a gene, which would be essential for such a test. We just detected linkage—the degree of gene sharing between related individuals with a known characteristic. Our experiments were designed to determine whether or not genes influence sexual orientation, not to test for the presence or effect of these genes on individuals. Our results were based purely on statistical measurements of a group and can say nothing about individual people.

Eventually, though, genes that influence sexual behavior probably will be isolated. Even when that day comes, however, it still won't be possible to conclusively test every person's current or future sexual orientation. We know this because even the identical twin of a gay man has a 50 percent or more chance of being heterosexual—even though he has the exact same genes and is reared by the same parents. In fact, the overall accuracy of any such test will probably be low because of the complex interplay of the genes with many different biological, environmental, social, cultural, and temporal factors—factors that cannot be measured or predicted by a blood test. Even with a test, parents could only be told the probability an unborn child would grow up to be straight or gay, a very weak statistic upon which to judge a human life.

No matter how inaccurate or impractical this hypothetical "genetic sexual orientation test" might be, someone may try to develop it, especially if there is a market for it. Gays and lesbians have reason to fear that such a test would be used against them, if the history of past "treatments," including castration, lobotomies, and psychotherapy, is any guide. Remember, though, that biology is neutral. There is no reason that gays, who, of course, can and do have children, could not use such a test to abort fetuses that had the "straight gene," or to refuse to hire people who didn't have a "gay" genetic profile.

I've yet to find a medical geneticist who would agree to perform such a test, but there is no guarantee it won't be developed by someone, somewhere. For that reason, it is important to consider three practical measures I (and others) could take to block such a test.

First, I can clearly and unambiguously oppose genetic "tests" or "therapies" for sexual orientation at every opportunity, including in professional journals, seminars, and the media.

Second, I can cooperate with groups developing guidelines and policies for the ethical use of genetic research. This includes government programs, such as the Ethical, Legal and Social Implications Initiative of the Human Genome Project, and private organizations such as the National Association of Gay and Lesbian Scientists and Technical Professionals, and Parents and Friends of Lesbians and Gays.

Third, I could try to use the law to withhold the "testing" technology, should it ever become available. Genetic testing as practiced in the United States requires commercialization, and commercialization generally requires protection of intellectual property through patents. If a lab does discover a "gay gene," it might be able to control the licensing of the technology.

This is not just a "minority" issue, or something that won't affect "normal" people. This is not just a question of sexual orientation, and as the Human Genome Project proceeds, genes for many human attributes probably will be discovered: genes that help make us smart or stupid, musical or tone deaf, skinny or fat. As each new genetic link is found, new ethical issues will arise. Do we want to have a society where parents can flip through a DNA catalogue and design their own "boutique baby"? Will we accept that it is perfectly reasonable to discriminate against people before they are born, or prevent them from being born, because we don't like their genes?

THE REAL DANGER

Although such hypothetical dangers are frightening, I believe there is a far greater danger. It's a danger that is present now, not in the future. A danger that is real, not hypothetical. A danger that is well recognized but thoroughly neglected. A danger that has cost human lives. The real danger is not studying sex at all.

Our society—and most others—doesn't just ignore the scientific study of sex, it actively discourages it. For example, the pioneering sex researcher Alfred Kinsey was vehemently attacked by politicians, religious leaders, and other scientists. As a result, he lost his funding from the Rockefeller Foundation and died a broken man. Even today, scientists find it difficult to obtain funding for anything remotely connected to sexuality.

Some people try to divide sexuality research into areas that are "useful"—those that directly affect health and reproduction—and those that are "dangerous," meaning those that seek to understand the "whys" as well as the "whats" of sexual attitudes, thoughts, and behaviors. This may work in the press or in Congress, but it doesn't work in real life. We can't teach people about sexuality if we don't understand what it is or how it develops. We'll never be able to modify unhealthy sexual practices if we don't understand why they are unhealthy and why people want to do them. And we won't have the scientists or educators to do this work until we have the funding for integrated and comprehensive research programs investigating all areas of human sexuality.

The neglect and discouragement of sexuality research has had devastating consequences. When AIDS swept across the United States in the late 1970s and early 1980s, it was almost exclusively restricted to gay men. The first thing the epidemiologists needed to know was "How many gay men are there in the United States?" The answer was "We don't know." Nobody has studied that since Kinsey, the scientists said, and look what happened to him.

When it became clear that AIDS was being sexually transmitted, the next important question was "What do gay men do in bed?" The answer was "We don't know, and who would ever pay for that kind of research?"

Once it was established that the primary route for HIV transmission was through anal intercourse, more practical and essential questions arose. How important is anal intercourse to gay men? Should safer-sex education stress no intercourse at all, or the use of condoms? Still there were no answers and no support for finding them.

Every time a scientist or a doctor said, "We don't know," more lives were lost. Many of these men didn't have to die. We do know—from the experience in countries like Australia where safer sex has been preached since the beginnings of the epidemic—that research and education make a difference.

What we don't know is what will be next and who it will strike. Will it be breast cancer in lesbians or prostate cancer in gay men? Will it be a new sexual disease or dysfunction that strikes only heterosexuals—or the young, or the old? Or will it just be more of the same: an unrelenting epidemic of unwanted teenage pregnancies, rape, abuse, and sexually transmitted diseases?

If we repeat the errors of the past, we won't know the answers until it's too late. To prevent that from happening we need sex education. We need a media campaign for safe sex—a campaign that isn't afraid of condoms. We need labs and funding and an infrastructure for sexuality research. The scientific study of sex can teach us more about how genes and hormones and the environment shape our desire. It can teach us how the mind works and what it means to be human. And above all it can keep us healthy, productive, and alive. Although the topic will probably always be controversial, when it comes to sexuality, we have far more to fear from ignorance than from knowledge.

APPENDIX A

A Linkage Between DNA Markers on the X Chromosome and Male Sexual Orientation

Dean H. Hamer, Stella Hu, Victoria L. Magnuson, Nan Hu, Angela M. L. Pattatucci

The role of genetics in male sexual orientation was investigated by pedigree and linkage analyses on 114 families of homosexual men. Increased rates of same-sex orientation were found in the maternal uncles and male cousins of these subjects, but not in their fathers or paternal relatives, suggesting the possibility of sex-linked transmission in a portion of the population. DNA linkage analysis of a selected group of 40 families in which there were two gay brothers and no indication of nonmaternal transmission revealed a correlation between homosexual orientation and the inheritance of polymorphic markers on the X chromosome in approximately 64 percent of the sib-pairs tested. The linkage to markers on Xq28, the subtelomeric region of the long arm of the sex chromosome, had a multipoint lod score of 4.0 ($P = 10^{-5}$), indicating a statistical confidence level of more than 99 percent that at least one subtype of male sexual orientation is genetically influenced.

Human sexual orientation is variable. Although most people exhibit a heterosexual preference for members of the opposite sex, a significant minority display a homosexual orientation. This naturally occurring variation presents an opportunity to explore the mechanisms underlying human sexual development and differentiation.

The role of genetics in sexual orientation has been previously approached by twin, adoption, and nuclear family studies. From the rates of homosexuality observed in the monozygotic and dizygotic twins, ordinary siblings, and adoptive (adopted in) brothers and sisters of homosexual men (*1*, *2*) and women (*3*, *4*), overall heritabilities of 31 to 74 percent for males and 27 to 76 percent for females were estimated. However, the precise extent of genetic loading is unclear because systematic data on relatives raised apart (adopted out) are not available and because the number and nature of the

The authors are with the Laboratory of Biochemistry, National Cancer Institute, National Institutes of Health, Bethesda, MD 20892.

putative inherited factors are unknown. The observation that male homosexuals usually have more gay brothers than gay sisters, whereas lesbians have more gay sisters than gay brothers, suggests that the factors responsible for this familial aggregation are at least partially distinct in men compared to women (*3, 5*).

Recent neuroanatomical studies have revealed differences between heterosexual and homosexual men in the structure of three regions of the brain; namely, the third interstitial nucleus of the anterior hypothalamus (*6*), the anterior commissure (*7*), and the suprachiasmatic nucleus (*8*). The role of gonadal steroids in the sexual differentiation of the mammalian brain is well established (*9*), but thus far the role of hormonal variations in normal human sexual development is unknown (*10*). Nonbiological sources of variation in human sexual expression have been under consideration in diverse disciplines including psychiatry, psychology, religion, history, and anthropology (*11*).

The goal of our work was to determine whether or not male sexual orientation is genetically influenced. We used the standard techniques of modern human genetics, namely pedigree analysis and family DNA linkage studies. Recent advances in human genome analysis, in particular the development of chromosomal genetic maps that are densely populated with highly polymorphic markers, make it feasible to apply such methods to complex traits, such as sexual orientation, even if these traits are influenced by multiple genes or environmental or experiential factors, or some combination of these (*12*). Our data indicate a statistically significant correlation between the inheritance of genetic markers on chromosomal region Xq28 and sexual orientation in a selected group of homosexual males.

Characteristics of study participants. The subjects studied were self-acknowledged homosexual men and their relatives over age 18. The initial sample for pedigree analysis consisted of 76 index subjects who were recruited through the outpatient HIV clinic at the National Institutes of Health Clinical Center, the Whitman-Walker Clinic in Washington, D.C., and local homophile organizations. One or more relatives from 26 of these families also participated in the project (total $n = 122$). The sample for the sib-pair pedigree study consisted of 38 pairs of homosexual brothers, together with their parents or other relatives when available, who were recruited through advertisements in local and national homophile publications. Two additional families who were originally in the randomly ascertained pool were added to this group for the DNA linkage study (total $n = 114$). Subjects signed an Informed Consent, approved by the NCI Clinical Review Subpanel, prior to donating blood and completing an interview or questionnaire covering childhood gender identification, childhood and adolescent sexual development, adult sexual behavior, the Kinsey scales, handedness, alcohol and substance use, mental health history, medical genetics screen, HIV status, and demographics (*13, 14*). The participants were white non-Hispanic (92 percent), African American (4 percent), Hispanic (3 percent), and Asian (1 percent) and had an average educational level of 15.5 $\pm$ 2.4 (mean $\pm$ SD) years and an average age of 36 $\pm$ 9 (mean $\pm$ SD) years.

Sexual orientation was assessed by the Kinsey scales, which range from 0 for exclusive heterosexuality to 6 for exclusive homosexuality (*13*). Subjects rated themselves on four aspects of their sexuality: self-identification, attraction, fantasy, and behavior. Of the homosexual subjects, >90 percent self-identified as either Kinsey 5 or 6 whereas >90 percent of their nonhomosexual male relatives self-identified as either 0 or 1 (Fig. 1). The sexual attraction and fantasy scales gave even greater dispersions between the groups, with ≥95 percent of the participants either less than Kinsey 2 or more than Kinsey 4. Only the sexual be-

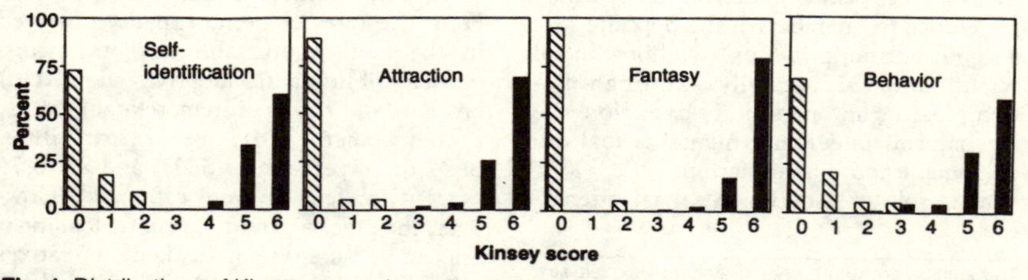

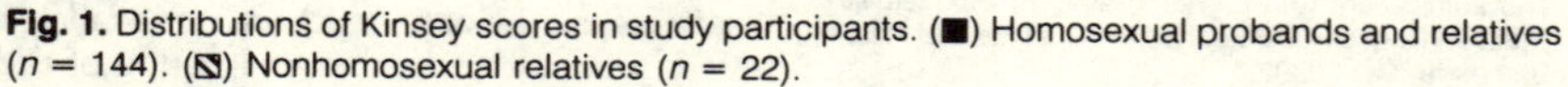
Fig. 1. Distributions of Kinsey scores in study participants. (■) Homosexual probands and relatives (n = 144). (▧) Nonhomosexual relatives (n = 22).

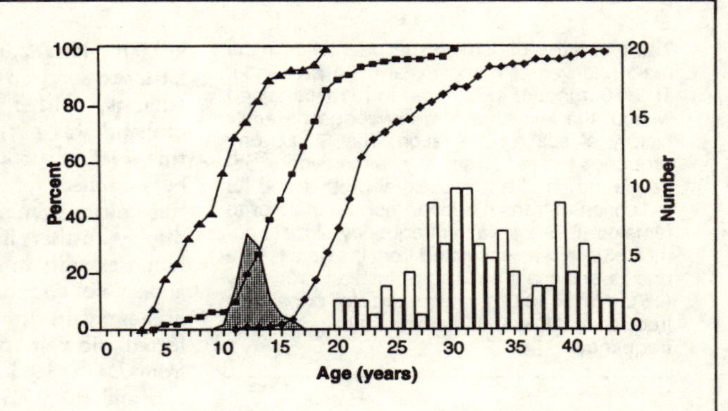

Fig. 2. Age of phenotypic expression for homosexual study participants. (▲) Age of first same-sex attraction, cumulative percent. (■) Age of self-acknowledgement, cumulative percent. (◆) Age of acknowledgement to others, cumulative percent. (▩) Age of puberty, percent. (□) Age of participants, number at each age. There were an additional 25 participants over age 44.

havior scale gave a small overlap between the two groups largely because of adolescent and early adult experiences. Therefore, for our study, it was appropriate to treat sexual orientation as a dimorphic rather than as a continuously variable trait. Similar bimodal distributions of Kinsey scores in males have been reported by others (*1, 2*).

The age of phenotypic expression of homosexuality was assessed by asking the subjects at what age they were first attracted to another male, when they acknowledged their sexual orientation to themselves, and when they acknowledged their orientation to others. Most of the subjects experienced their first same-sex attraction by age 10, which was prior to the average age of puberty at 12 years (Fig. 2). Self-acknowledgement occurred over a broad range of ages between 5 and 30 years, with the greatest increase occurring between years 11 and 19. The mean age for public acknowledgement was 21 years, which is similar to the average age for "coming out" reported by others (*15*). Since the average age of our subjects was 36 ± 9 years, we did not correct for age-dependent phenotypic expression in subsequent analyses.

Pedigree analysis. Traits that are genetically influenced aggregate in families and, in the case of dominant or sex-linked inheritance, are transmitted from one generation to the next. Family histories were collected from 114 homosexual male probands who were asked to rate their fathers, sons, brothers, uncles, and male cousins as either definitely homosexual (Kinsey 5 or 6, acknowledged to the proband or another family member) or not definitely known to be homosexual (heterosexual, bisexual, or unclear). The reliability of the probands' assessment of their family members' sexual orientation was estimated by conducting interviews with 99 relatives of the index subjects. All (69/69) of the relatives identified as definitely homosexual verified the initial assessment, as did most (27/30) of the relatives considered to be nonhomosexual; the only possible discrepancies were one individual who considered himself to be asexual and two subjects who declined to answer all of the interview questions. Hence describing individuals as either homosexual or nonhomosexual, while undoubtedly overly simplistic, appears to represent a reliable categorization of the population under study.

On the basis of a separate study in which the uncles and male cousins of lesbians were interviewed (*16*), we estimated that the population prevalence of male homosexuality is 2 percent (14/717). Although this rate is lower than the popularly accepted figures of 4 to 10 percent for male homosexuality, probably due to the more stringent definition applied here, it was considered more accurate for this analysis since the sampling, interview format, and definition of homosexual orientation were identical to those used in the male study. Similarly low rates for the population incidence of homosexuality have been reported when recent sexual behavior was used as the criterion (*17*).

The pedigree analysis for the male relatives of the 76 randomly ascertained homosexual male probands indicated (Table 1) that the highest rate of homosexual orientation was in brothers, who had a 13.5 percent chance of being gay, representing a significant 6.7-fold increase over the estimated background rate of 2 percent ($P < 0.001$). Among more distant relatives, only two groups had significantly higher rates of homosexual orientation than the population incidence, namely maternal uncles and the sons of maternal aunts. Both of these maternally related classes of relatives had rates of ≈7.5 percent, which were significantly higher than the background rate ($P < 0.01$). By contrast, fathers and all other types of paternally related relatives had rates that were lower or not significantly

Table 1. Rates of homosexual orientation in the male relatives of homosexual male probands. The 76 random probands were ascertained without the investigator's knowledge of family history of sexual orientation. The 38 sib-pair probands were selectively ascertained because they had a homosexual brother and no indication of transmission through fathers or to females. The population frequency of male homosexuality was estimated from the data for the uncles and male cousins of lesbian probands (*16*). $^{**}P < 0.001$ compared to population frequency. $^{*}P < 0.01$ compared to population frequency.

Relationship	Homosexual/ total	Percent
Random probands (n = *76*)		
Father	0/76	0
Son	0/6	0
Brother	14/104	13.5**
Maternal uncle	7/96	7.3*
Paternal uncle	2/119	1.7
Maternal cousin, aunt's son	4/52	7.7*
Maternal cousin, uncle's son	2/51	3.9
Paternal cousin, aunt's son	3/84	3.6
Paternal cousin, uncle's son	3/56	5.4
Sib-pair probands (n = *38*)		
Maternal uncle	6/58	10.3**
Paternal uncle	1/66	1.5
Maternal cousin, aunt's son	8/62	12.9**
Maternal cousin, uncle's son	0/43	0
Paternal cousin, aunt's son	0/69	0
Paternal cousin, uncle's son	5/93	5.4
Population frequency		
Uncles and cousins of female probands	14/717	2.0

different from the background. Background rates of homosexuality were also observed in the female relatives of the homosexual male probands (except for sisters, who had a 5.4 percent rate versus a 1 percent background rate) and in the male relatives of lesbian probands (except brothers, who had a 4.7 percent rate) (*16*).

Although the observed rates of homosexual orientation in the maternally derived uncles and male cousins of gay men were higher than in female and paternally related male relatives, they were lower than would be expected for a simple Mendelian trait. Furthermore, there was a substantial number of families in which lesbians or paternally related gay men were present. This could be explained if some instances of homosexuality were male-limited and maternally inherited whereas others were either sporadic, not sex-limited, or not maternally transmitted. To test this, we recruited 38 families in which there were two homosexual brothers, no more than one lesbian relative, and no indication of direct father-to-son transmission of homosexuality (that is, neither the father nor son of a proband was gay). We hypothesized that this selected population of families would be enriched for the putative maternally transmitted genetic factor and therefore display further increases in the rates of homosexuality in maternally derived uncles and male cousins. Indeed, the rates of homosexuality in the relatives of these selected sib-pair probands were increased from 7.3 to 10.3 percent for maternal uncles and from 7.7 to 12.9 percent for the sons of maternal aunts (Table 1). By contrast, the rates of homosexuality in the other types of male relatives were unchanged or decreased compared to the initial study. The differences between the random and sib-pair populations were not significantly different ($P > 0.1$); however, the differences between all maternal relatives as compared to all nonmaternal relatives were significant within both the randomly ascertained group ($P < 0.05$) and the sib-pair group ($P < 0.001$).

Several examples of the apparent maternal transmission of male homosexual orientation are shown in Fig. 3. Families DH99002 and DH99017, which were randomly ascertained, are characterized by a single gay man in each of three maternally related generations. In family DH321, which was recruited as part of the sib-pair study, a pair of homosexual brothers have a maternally related gay nephew and uncle. Family DH210, which was ascertained as part of a separate study, contains seven homosexual males, all related through the sequential marriage of two sisters to the same husband in generation II. In several families, maternally related half-brothers or half-cousins shared a homosexual orientation (*16*). The striking feature of these multiplex pedigrees is the absence of transmission through the paternal line and the paucity of female homosexuals.

These results demonstrate increased rates of homosexual orientation not only in the brothers of gay men, as has been previously reported (*1, 2*), but also in maternal uncles and the sons of maternal aunts (*18*). Because uncles and cousins share inherited information with the index subjects, but are raised in different households by different parents, this observation favored an interpretation based on genetics rather than the rearing environment and suggested that linkage studies might be fruitful.

X chromosome linkage. One explanation for the maternal transmission of a

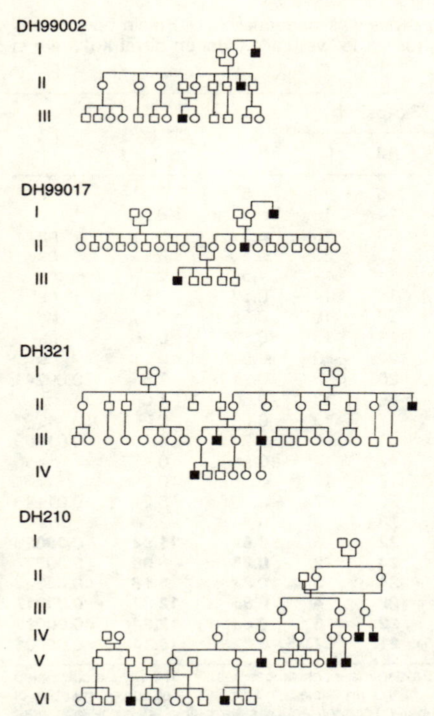

Fig. 3. Family pedigrees displaying apparent maternal transmission of male homosexuality. Families DH99002 and DH99017 were from the randomly ascertained group of probands. Families DH321 and DH210 were selected because many members were homosexual. (■) Homosexual males. (□) Nonhomosexual males. (○) Nonhomosexual females.

male-limited trait is X chromosome linkage. Since males receive their single X chromosome exclusively from their mothers, any trait that is influenced by an X-linked gene will be preferentially passed through the mother's side of the family. DNA linkage analysis provides the means to distinguish X-linked inheritance from competing hypotheses such as maternal effects, imprinting, decreased reproductive rates of expressing males, or differential knowledge concerning maternal versus paternal family members. If the X chromosome contains a gene that increases the probability of an individual's being homosexual, then genetically related gay men should share X chromosome markers close to that gene. If no such gene exists, then no statistically significant correlations between sexual orientation and X chromosome markers will be observed (*19*).

We performed the linkage analysis on the selected population of families described above in which there were two homosexual brothers. This sib-pair experimental design has several theoretical and practical benefits (*20*): (i) it is nonparametric and independent of gene penetrance and frequency; (ii) it is capable of detecting a single linked locus even if additional genes or environmental conditions are required to express the trait; (iii) it is more powerful to study siblings than more distant relatives for traits displaying limited familiality; (iv) "false negatives" (individuals who have or will have a homosexual orientation but choose to identify themselves as heterosexual) are irrelevant to the analysis because they are not studied; (v) "false positives" (individuals who have a heterosexual orientation but choose to identify themselves as homosexual) are expected to be rare; (vi) the sib-pair method is more stable to errors in genotyping and to mistakes or alterations in phenotype than are large pedigree methods; and (vii) it was more practical to obtain the cooperation of nuclear sib-pair families than of multigenerational families.

The sample for the linkage analysis consisted of 40 pairs of homosexual brothers (38 from the sib-pair pedigree study and 2 from the random sample) together with their mothers or other siblings if available. DNA was prepared from all available members of these families and typed for a series of 22 markers that span the X chromosome. Each sib-pair was scored as either concordant-by-descent (D) if the mother was known to be heterozygous and both sons inherited the same allele, concordant-by-state (S) if the mother was unavailable and both sons shared the same allele, discordant (−) if the two sons carried different alleles, or noninformative (n) if the mother was homozygous for the marker. For families in which DNA from the mother was not available, the data for the concordant-by-state pairs were corrected for the possibility that the mother was homozygous for the marker by taking into account the population frequency of the allele coinherited by the two sons (*19, 20*). Using a likelihood ratio test, we then calculated for each locus the probability (z_1) of the brothers sharing the marker by-descent and the statistical significance (P) of deviations from the value of $z_1 = 1/2$ expected under the null hypothesis of no linkage.

The X chromosome markers used for linkage analysis were simple sequence repeats, variable number of tandem repeats, and restriction fragment length polymorphisms, all of which were detected by the polymerase chain reaction (PCR) (Table

Table 2. Summary of linkage results. Linkage analysis was performed on 40 male homosexual sib-pairs; 22 X chromosome markers were used (*30*). The five marker loci on distal Xq28 are in boldface.

Locus	Location	AL*	HET†	Sib-pairs‡			z_1§	$2\ln L(z_1)$‖	P¶
				[D]	[S]	[–]			
A. .KAL	p22	6	0.77	5	16	14	0.51	0.01	ns
B. .DXS996	p22	11	0.84	7	14	18	≤.5	≤0	ns
C. .DXS992	p	8	0.87	6	13	19	≤.5	≤0	ns
D. .DMD1	p21	9	0.78	3	10	23	≤.5	≤0	ns
E. .DXS993	p11	6	0.80	3	14	17	≤.5	≤0	ns
F. .DXS991	p	8	0.77	8	14	14	0.57	0.61	ns
G. .DXS986	q	10	0.71	7	20	10	0.65	2.11	ns
H. .DXS990	q	7	0.76	4	19	13	0.55	0.25	ns
I. .DXS1105	q	5	0.48	3	20	9	≤.5	≤0	ns
J. .DXS456	q21	10	0.85	8	20	8	0.75	7.95	0.00241
K. .DXS1001	q26	10	0.82	8	16	13	0.60	1.09	ns
L. .DXS994	q26	5	0.75	7	17	13	0.55	0.26	ns
M. .DXS297	q27	5	0.70	5	21	8	0.71	4.25	0.01963
N. .FMR	q27	17	0.79	6	17	14	0.56	0.45	ns
O. .FRAXA	q27	8	0.72	4	17	13	0.56	0.38	ns
P. .DXS548	q27	6	0.67	7	20	7	0.73	5.21	0.01123
Q. .GABRA3	q28	4	0.35	2	23	3	0.74	2.39	ns
R. .DXS52	**q28**	**12**	**0.79**	**9**	**22**	**6**	**0.81**	**11.83**	**0.00029**
S. .G6PD	**q28**	**2**	**0.36**	**4**	**24**	**2**	**0.85**	**6.38**	**0.00577**
T. .F8C	**q28**	**2**	**0.41**	**5**	**24**	**3**	**0.82**	**6.56**	**0.00522**
U. .DXS1108	**q28**	**6**	**0.71**	**8**	**22**	**4**	**0.85**	**12.87**	**0.00017**
V. .DXYS154#	**q28**	**10**	**0.71**	**8**	**22**	**5**	**0.83**	**12.84**	**0.00017**
R/S/T/U/V	**q28**		**0.99**	**12**	**21**	**7**	**0.82**	**18.14**	**0.00001**

*AL is the number of different alleles observed in 62 to 150 independent chromosomes. †HET is the calculated heterozygosity; HET = $1 - \Sigma f_i^2$, where f_i = frequency of the *i*th allele. ‡[D] is the observed number of concordant-by-descent pairs; [S] is the observed number of concordant-by-state pairs; [–] is the observed number of discordant pairs; noninformative pairs are not included in this analysis. §z_1 is the estimated probability that two homosexual brothers share the marker locus by-descent (*31*). ‖$L(z_1)$ is the ratio of the likelihoods of the observed data at z_1 versus the null hypothesis of $z_1 = 1/2$ (*31*). ¶*P* (one-sided) was calculated by taking $2\ln L(z_1)$ to be distributed as a chi-squared statistic at one degree of freedom; ns: $P > 0.05$. #Only the maternal, X-linked contribution was considered for this sex-linked locus (*23*).

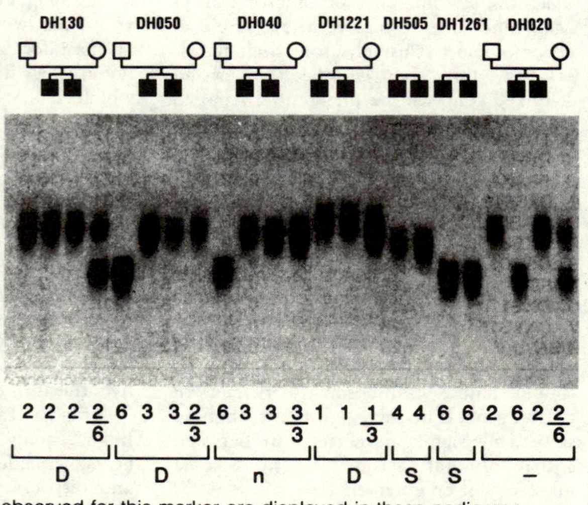

Fig. 4. Linkage analysis of the $(GT)_nGC(GT)_n$-repeat locus DXS1108. Genotypes were determined by PCR amplification with the use of the DXS1108 [SDF2] primers (*21*) in the presence of [α-^{32}P]dCTP. A portion from each reaction was electrophoresed through a 6 percent denaturing acrylamide gel and exposed to x-ray film for 2 hours (*30*). The top diagrams show the family pedigrees (same symbols as in Fig. 3). The lower numbers show the allele assignments and the determination of status as concordant-by-descent (D), noninformative (n), concordant-by-state (S), or discordant (–). Five out of the six alleles observed for this marker are displayed in these pedigrees.

2). Heterozygosities, which were determined by analyzing 62 to 150 independent X chromosomes from the sib-pair and related populations, ranged from 0.35 to 0.87. An example of genotype determination with a $(GT)_nGC(GT)_n$-repeat marker, DXS1108, is shown in Fig. 4. Despite the presence of shadow bands, the individual alleles were readily distinguishable, and concordant and discordant sib-pairs could be clearly differentiated. As expected for this X-specific marker, the alleles inherited by the sons were derived exclusively from the mother. By contrast, the marker DXYS154, which lies on the tip of Xq in a region of subtelomeric homology and genetic exchange between the X and Y chromosomes, displayed alleles contributed by both the father and the mother (*21*). As expected for this tightly sex-linked marker, almost all of the male siblings inherited the same Y chromosome allele from their fathers (*22*); therefore, only the contribution from the maternal X chromosome was considered in the analysis of this locus (*23*).

The linkage analysis included a statistical analysis of the pair-by-pair data (Tables 2 and 3) and multipoint mapping analysis of the X chromosome (Fig. 5). The main outcome was the detection of linkage between homosexual orientation and markers in the distal portion of Xq28. Each of the five markers in this region gave values of $z_1 > 0.8$, and for the three most heterozygous loci the data were significant at $P < 0.0003$ (Table 2). The five terminal loci on Xq28 are clustered within 2.8 to 4.3 cM (*21, 24*), and within our collection of 40 families exhibited no unequivocal intramarker recombination events (*25*). Therefore, the entire distal region of Xq28 could be considered as a single extended locus with a haplotype heterozygosity of 0.99; this transformation of the data increases the power to detect linkage by decreasing the uncertainties due to nongenotyped and noninformative mothers. Of the 40 sib-pairs, 33 were concordant for all markers within this region, whereas 7 pairs were discordant at one or more loci (Table 3). This analysis gives a value of $z_1 = 0.82$ at a significance of $P = 1.2 \times 10^{-5}$.

Evaluation of the data by multipoint mapping with the LINKMAP routine of the computer program LINKAGE 5.1 supported the linkage between homosexual orientation and distal Xq28. The model used for analysis was an X-linked, male-specific gene with a mutation rate of 0. The population frequency of the homosexuality-associated allele was assumed to be 0.02, and penetrances were set at 0 for all females, 0 for males lacking the trait-associated allele, and 0.5 for males having the trait-associated allele; heterosexual brothers were not included in the analysis. The peak multipoint lod score was 3.96 to 4.02 (Fig. 5), depending on whether compressed or full allele information was used; because the lod score is the logarithm to the base 10 of the odds ratio, this corresponds to an odds ratio of ≈10,000:1. The apparent location of the peak was 8 cM distal of DXYS154. However, this is likely to be an overestimate due to the well-known phenomenon

Table 3. Pair-by-pair linkage analysis. The results of the X chromosome linkage analysis for each homosexual male sib-pair family are shown. The marker loci are described in Table 2. Each sib-pair was scored as concordant-by descent (D), concordant-by-state (S), discordant (–), or noninformative (n). The first 33 families are concordant for the distal portion of Xq28 (loci R-S, boldface); the last 7 are discordant.

Proband ID #	Marker locus ABCDEFGHIJKLMNOPQRSTUV
DH130	nDDDDDnDnDDnnDnDnnnDDD
DH050	--D-n-nnnDnDnnnnnDnnDD
DH040	n--nDDDnDnDnDDDDnDDnnD
DH371	DD--n-DDnDDDnnnnnDnnnD
DH1221	-----D--DDnDDDDDDnDDDD
DH070	DDD--nD-DDDDnDnDnDDDDn
DH471	---D----nn--DDDDDDnnDD
DH060	nDn-------n-DDDnDnDnDDD
DH151	DnDnDDn-----DnDDnDnnnD
DH170	n---nDDDnDDn---nnDDDDn
DH441	DD---n--nnD-n--nnDnnnn
DH301	DDDSSDSDSDS-----SSSSDS
DH411	SSSSSSSSSSSSSSSSSSSSSS
DH1281	S---SSSSSSSSSSSSSSSSSS
DH231	----SSSSSS-SSSSSSSSSSS
DH1261	----SSSS-S--SSSSSSSSSS
DH1391	SS-S-SSSSSSSSSSSSSSSSS
DH331	-----SSSSSSSSSSSSSSSSS
DH421	SSS--SSSSSSSSSSSSSSSSS
DH1131	SS--S-SSSSSSSSSSSSSSSS
DH311	-SSSS-SSSSSSSSSSSSSSSS
DH1081	SSS---SSSSS-SSSSSSSSSS
DH321	SS--S--SSSS-SSSSSSSSSS
DH505	SSS-SS-S---SSSSSSSSSSS
DH1171	--SSS-S--S-SSSSSSSSSSS
DH281	---------S-SSSSSSSSSSS
DH461	--SS----S--SSSSSSSSSSS
DH1191	SSS----SS---SSSSSSSSSS
DH1381	SS---SS-SS-S---SSSSSSS
DH1331	SS----S-S-S---S-SSSSSS
DH1211	---S-SSSSSSS----SSSSSS
DH1061	SSS-SSSSSS------SSSSSS
DH1101	S-SSSSSSSSSSS----SSSSS
DH391	n-nnnn-nnDDD--n-n-nnnn
DH220	-----nDD----n--nn-n--n
DH020	--DDnDD---DD---nnn-n--
DH1041	SDDnnDDnnnn-S--S--nSn-
DH1121	SSSSSSSS-SSSS-SS-SSS-
DH1021	--SSSSSSSSSS--SS-----
DH1051	SS----S--SS-S-----S---

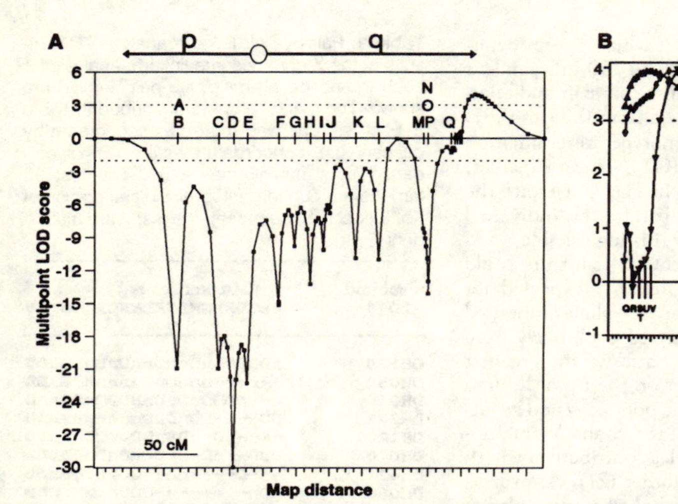

Fig. 5. (**A**) Multipoint mapping of the sexual orientation–related locus on the X chromosome. The data from the linkage study [Table 2; (*30*)] were analyzed by the LINKMAP routine of LINKAGE 5.1 implemented on a VAX 6620 computer at the Biomedical Supercomputing Center, Frederick Cancer Research and Development Center. The analysis was done with compressed alleles on five overlapping sets of six fixed loci compared to one test locus. Five recombination fractions between each pair of loci were evaluated. The program parameters were as follows: population frequency of trait-associated allele = 0.02, penetrance for all females = 0, penetrance for males lacking the trait-associated allele = 0, penetrance for males having the trait-associated allele = 0.5, and mutation rate = 0. The maximum lod score was not significantly altered by varying the penetrance for males having the trait-associated allele between 0.05 and 1, by setting the mutation rate to 0.01, nor by changing the distances between the fixed marker loci. (**B**) The data for Xq28 were evaluated with the use of the full allele information under the following models: ▼---▼, frequency of trait-associated allele = 0.02, penetrance of non-trait–associated allele = 0 (standard conditions); ◆---◆, frequency of trait-associated allele = 0.1, penetrance of non-trait–associated allele = 0; ▲---▲, frequency of trait-associated allele = 0.02, penetrance of non-trait–associated allele = 0.05.

of biased recombination fraction estimation in the case of complex traits where the analysis model differs from the true model (*19*). Therefore, the data were reanalyzed under two alternative models (Fig. 5B). When the frequency of the trait-associated allele was increased from 0.02 to 0.1, as suggested by Risch and Giuffra (*26*), the peak lod score decreased to 3.9 and the distance from DXYS154 decreased to 5 cM. When the penetrance of the non-trait–associated allele was increased from 0 to 0.05, giving a substantial level of phenocopies, the peak lod score of 3.9 fell directly over DXYS154, and the lod scores throughout distal Xq28 were greater than 3.5. Given that DXYS154 lies within 1 Mb of the telomere (*27*), these latter models probably yield more accurate estimates of the locus position. More precise mapping will require more distal markers, a larger number of families, and additional information concerning the trait parameters.

There was no significant evidence for linkage between sexual orientation and loci lying outside of Xq28. Most of the markers on the remainder of the long arm, and all of the markers on the short arm, gave values of z_1 that were statistically indistinguishable from the null hypothesis ($P > 0.05$) (Table 1). Although there was a moderate excess of concordant pairs at the markers DXS456, DXS297, and DXS548 ($0.002 \leq P \leq 0.02$), it is unlikely that these loci play a significant role in sexual orientation because they were adjacent to markers that gave negative results. Furthermore, multipoint mapping gave lod score less than −2 throughout the region between the KAL locus at Xp22.3 and the DXS994 locus at Xq26 and around the fragile X locus at Xq27.3. However, a much larger sample would be required to stringently eliminate these regions from playing a role in sexual development in a small proportion of families.

Contribution of genetics to male sexual orientation. The proof for the involvement of genes in a human behavioral trait must ultimately consist of the chromosomal mapping of the loci and isolation of the relevant DNA sequences. Such molecular studies are essential to separate the role of inheritance from environmental, experiential, social, and cultural factors. DNA linkage studies of families in which the trait appears to be genetically segregating represent the first step in this approach.

We have now produced evidence that one form of male homosexuality is preferentially transmitted through the maternal side and is genetically linked to chromosomal region Xq28. In a selected population of families in which there were two homosexual brothers and no transmission through fathers or to females, 33 of 40 sib-pairs had coinherited genetic information in this subtelomeric region. Observing such an association by chance alone has a type I error rate of approximately 0.001 percent for a single tested region of the genome (haplotype $P = 1.2 \times 10^{-5}$), and therefore an error rate of less than 0.03 percent for a collection of 22 independent markers ($P = 22 \times 1.2 \times 10^{-5} = 0.0003$). Similarly, multipoint linkage mapping gave a peak lod score of 4.0, which is associated with an overall type I error level of 0.5 percent, even for a complete genome search (*12, 19*). Thus, both forms of analysis indicate that the linkage results are statistically significant at a confidence level of >99 percent. As with all linkage studies, replication and confirmation of our results are essential. The observed excess coinheritance of Xq28 markers by homosexual brothers is not due to segregation distortion because normal, Mendelian segregation has been demonstrated for many different Xq28-linked traits and polymorphic markers (*28*). Rather, it appears that Xq28 contains a gene that contributes to homosexual orientation in males.

There were seven pairs of brothers who did not coinherit all of the Xq28 markers. Such discordant pairs could arise because of homozygosity of the mother at the sexual orientation–related locus, recombination between the locus and a marker gene, genetic heterogeneity, or nongenetic sources of variation in sexual orientation. We estimate that the last two categories comprise approximately 36 percent of the sib-pair population (*29*). At present, we can say nothing about the fraction of all instances of male homosexuality that are related or unrelated to the Xq28 candidate locus because of the selection for genetically loaded families that is imposed by linkage methods. We also have no information about the role, or lack thereof, of the Xq28 region in multiplex families containing multiple gay men or lesbians (or both) (*29*), nor about the presence or absence of the homosexuality-associated allele in brothers or other male relatives who identify as heterosexual. Given the overall complexity of human sexuality, it is not surprising that a single genetic locus does not account for all of the observed variability. Sib-pairs that are discordant at Xq28 should provide a useful resource for identifying additional genes or environmental, experiential, or cultural factors (or some combination of these) that influence the development of male sexual orientation.

Our experiments suggest that a locus (or loci) related to sexual orientation lies within approximately 4 million base pairs of DNA on the tip of the long arm of the X chromosome. Although this represents less than 0.2 percent of the human genome, it is large enough to contain several hundred genes. The fine mapping and eventual isolation of this locus will require either large numbers of sib-pairs, more extended families, or the complete DNA sequence of the region. Once a specific gene has been identified, we can find out where and when it is expressed and how it ultimately contributes to the development of both homosexual and heterosexual orientation. The Xq28 region is characterized by a high density of genetic loci (*28*) and contains both repeated DNA sequences (*27*) and a pseudoautosomal region of homology and genetic exchange between the X and Y chromosomes (*21*). Recombination between tandemly repeated sequences, or between active and inactive loci on the X and Y chromosomes, could generate DNA sequence variants at a high rate and thereby account for the genetic transmission of a trait that may reduce reproduction.

The subjects for our linkage study were males who self-identified as predominantly or exclusively homosexual within the context of modern American society; such studies could be broadened to include individuals who identify as bisexual or ambisexual. The role of the Xq28 candidate locus, and of other chromosomal regions, in female sexual orientation remains to be tested. Although nuclear family studies suggest that the overall heritability of sexual orientation is similar in men and women (*2, 4*), their pedigree segregation patterns appear to be distinct (*16*).

Our work represents an early application of molecular linkage methods to a normal variation in human behavior. As the human genome project proceeds, it is likely that many such correlations will be discovered. We believe that it would be fundamentally unethical to use such information to try to assess or alter a person's current or future sexual orientation, either heterosexual or homosexual, or other normal attributes of human behavior. Rather, scientists, educators, policy-makers, and the public should work together to ensure that such research is used to benefit all members of society.

REFERENCES AND NOTES

1. R. C. Pillard and J. D. Weinrich, *Arch. Gen. Psychiatry* **43**, 808 (1986).
2. J. M. Bailey and R. C. Pillard, *ibid.* **48**, 1089 (1991).
3. J. M. Bailey and D. S. Benishay, *Am. J. Psychiatry* **150**, 272 (1993).
4. J. M. Bailey, R. C. Pillard, M. C. Neale, Y. Agyei, *Arch. Gen. Psychiatry* **50**, 217 (1993).
5. J. D. Weinrich, *Sexual Landscapes* (Scribner, New York, 1987); R. C. Pillard, in *Homosexuality/Heterosexuality: Concepts of Sexual Orientation*, D. McWhirter, S. A. Sanders, J. M. Reinisch, Eds. (Oxford Univ. Press, London, 1990), pp. 88–100; J. M. Bailey, L. Willermaan, C. Parks, *Arch. Sex. Behav.* **20**, 277 (1991).
6. S. LeVay, *Science* **253**, 1034 (1991).
7. L. S. Allen and R. A. Gorski, *Proc. Natl. Acad. Sci. U.S.A.* **89**, 7199 (1992).
8. D. F. Swaab and M. A. Hofman, *Brain Res.* **537**, 141 (1990).
9. R. A. Gorski, in *Brain Endocrinology*, M. Motta, Ed. (Raven, New York, 1991), pp. 71–104; B. S. McEwen *et al.*, *J. Steroid Biochem. Mol. Biol.* **39**, 223 (1991).
10. H. F. Meyer-Bahlburg, *Clin. Endocrinol. Metab.* **11**, 681 (1982).
11. R. Friedman, *Psychoanal. Rev.* **73**, 483 (1986); C. A. Tripp, *The Homosexual Matrix* (Meridian, New York, 1987); R. A. Isay, *Being Homosexual* (Avon, New York, 1989); J. S. Spong, *Living in Sin?* (Harper and Row, San Francisco, 1988); R. Schow, W. Schow, M. Raynes, Eds., *Peculiar People: Mormons and Same-Sex Orientation* (Signature, Salt Lake City, 1991); A. L. Rouse, *Homosexuals in History* (Dorset, New York, 1977); M. Dubermaan, M. Vicinas, G. Chauncey, Jr., Eds., *Hidden from History: Reclaiming the Gay and Lesbian Past* (Meridian, New York, 1989); F. L. Whitam and R. M. Mathy, *Male Homosexuality in Four Societies: Brazil, Guatemala, the Philippines, and the United States* (Praeger, New York, 1986).
12. E. S. Lander and D. Botstein, *Proc. Natl. Acad. Sci. U.S.A.* **83**, 7353 (1986); E. S. Lander, in *Genome Analysis*, K. E. Davis, Ed. (IRL Press, New York, 1988), pp. 171–189.
13. A. C. Kinsey and W. B. Pomeroy, *Sexual Behavior in the Human Male* (Sanders, Philadelphia, 1948).
14. K. Freund, R. Langevin, J. Satterberg, B. Steiner, *Arch. Sex Behav.* **6**, 507 (1977); N. Buhrich J. M. Bailey, N. G. Martin, *Behav. Genet.* **21**, 75 (1991).
15. A. P. Bell and M. S. Weinberg, *Homosexualities: A Study of Diversity among Men and Women* (Simon and Schuster, New York, 1978).
16. A. Pattatucci and D. Hamer, unpublished results.
17. ACSF Investigators, *Nature* **360**, 407 (1992); A. M. Johnson, J. Wadsworth, K. Wellings, S. Bradshaw, J. Field, *ibid.*, p. 410.
18. Increased rates of homosexuality in the maternal compared to paternal relatives of homosexual men can also be discerned in the data of G. W. Henry [*Sex Variants: A Study of Homosexual Patterns* (Hoeber, New York, 1941)], B. Zuger [*Arch. Sex. Behav.* **18**, 155 (1989)], and Pillard and Weinrich (*1*).
19. J. Ott, *Analysis of Human Genetic Linkage* (Johns Hopkins Univ. Press, Baltimore, 1991).
20. B. K. Suarez, J. Rice, T. Reich, *Ann. Hum. Genet.* **42**, 87 (1978); G. Thomson, *Am. J. Hum. Genet.* **39**, 207 (1986); W. Blackwelder and R. Elston, *Genet. Epidemiol.* **2**, 85 (1985); L. Goldin and E. Gershon, *ibid.* **5**, 35 (1988); N. Risch, *Am. J. Hum. Genet.* **46**, 222 (1990); *ibid.*, p. 242; D. T. Bishop and J. A. Williamson, *ibid.*, p. 254.
21. D. Freije, C. Helms, M. S. Watson, H. Donis-Keller, *Science* **258**, 1784 (1992).
22. The published male recombination rate between sex and DXYS154 is 4/195 ≈ 2 percent (*21*). The observed rate in our sib-pair population was 1/36 ≈ 3 percent. The single recombinant family was DH441, for which the genotypes (alleles) were mother = [6,6], son 1 = [2,6], and son 2 = [6,6]; therefore this family was noninformative for the maternal X chromosome.
23. The maternal contribution for DXYS154 was calculated as follows. Families of type mother (alleles) = [1,2], and son 1 = son 2 = [1,*x*] or [2,*x*], where *x* = any allele, were scored as concordant-by-descent. Families of type son 1 = son 2 = [1,1] were scored as concordant-by-state with an allele frequency of f_1. Families of type son 1 = son 2 = [1,2] were scored as concordant-by-state with an allele frequency of $(f_1^2 + f_2^2)/(2f_1 + 2f_2 - f_1^2 - f_2^2 - 2f_1f_2)$. Families of type son 1 = [1,2] and son 2 = [1,3] were scored as discordant. Families of type mother = [1,1] were noninformative. If the homosexuality-associated gene were in fact derived from the father rather than from the mother in some families, this treatment of the data would decrease rather than increase the evidence for linkage.
24. J. Weissenbach *et al.*, *Nature* **359**, 794 (1992); CEPH database, version 6.0.
25. There was one possible recombination event, in family DH1101, between the proximal Xq28 locus GABRA3, which lies approximately 2 cM centromeric of DXS52, and the distal Xq28 gene cluster (Table 3). However, this could not be confirmed because the mother was not available for genotyping.
26. N. Risch and L. Giuffra, *Hum. Hered.* **42**, 77 (1992).
27. D. Freije and D. Schlessinger, *Am. J. Hum. Genet.* **51**, 66 (1992).
28. There appears to be no published evidence for segregation distortion for several well-studied Xq28-linked traits such as color blindness and G6PD deficiency [V. A. McKusick, *Mendelian Inheritance in Man* (Johns Hopkins Univ. Press, Baltimore, 1992)]. Analysis of the CEPH database (version 6.0) for the Xq28 marker loci F8C and DXS52 also showed no indication of segregation distortion in 33 informative families containing 142 sons. Analysis of the first two sons in each family as a single pair gave 18 concordant-by-state pairs versus 15 discordant pairs ($z_1 = 0.55$, $2\ln L(z_1) = 0.3$, $P >> 0.05$). Analysis of all $n(n - 1)/2$ pairs for each family (where n is the number of sons) gave 152 concordant-by-state pairs compared to 162 discordant pairs ($z_1 \leq 0.5$, $P >> 0.05$).
29. The fraction (α) of a sib-pair population in which a trait is associated with excess coinheritance of an X-linked marker is minimally estimated by $\alpha = 2z_1 - 1$, which takes into account the fact that 1/2 of all sib-pairs will coinherit the marker by chance alone. Therefore the fraction of the sib-pair population in which the trait is *not* linked to the gene is $1 - \alpha = 2(1 - z_1)$. For Xq28, $z_1 \approx 33/40 = 0.82$, giving $1 - \alpha = 0.36 = 36$ percent. The proportion

of the entire population in which the trait is linked to a marker cannot be estimated without (i) information on the frequencies and penetrances of the linked trait allele and additional loci and (ii) the frequency of phenocopies, none of which are known; under the simple model of a single gene and a high rate of phenocopies, Xq28 could account for as little as 10 percent of total variance. The contribution of Xq28 to sib-pairs cannot be extrapolated to larger families without further information. We have observed considerably greater variability in sexual development and expression in families containing more than two gay brothers or multiple lesbians (or both), suggesting that ascertainment may be more complicated in these cases.

30. Of the 40 families, there were 14 for which DNA from the mother was genotyped for all loci (DH130, 050, 040, 371, 1221, 070, 471, 060, 151, 170, 441, 391, 220, and 020), 1 family for which DNA from the mother was genotyped for some but not all loci (DH1041), and 1 family for which DNA from a sister but not the mother was genotyped (DH301). For the remaining 24 families, DNA was available only from the two brothers. DNA was prepared from peripheral blood by SDS lysis and salt precipitation [D. Lahiri and J. L. Nurnberger, Jr., *Nucleic Acids Res.* **19**, 5444 (1991)]. The PCR procedure followed standard conditions for each primer pair, with analysis by electrophoresis on 6 percent denaturing acrylamide, 8 percent acrylamide, or 1 percent agarose gels. Marker loci B, C, E, F, G, H, I, K, and L were as described (*24*); marker loci U and V were as described (*21*). Marker loci A (primers Kal.PCR1.1/Kal.PCR1.2), J (primers XG30BL/XG30BR), M (primers VK23F/VK23R), and P (primers RS46-CA1/RS46-CA2) were as described in the Genome Data Base (William H. Welch Medical Library, Johns Hopkins University). Marker locus D (DMD1) was described by J. P. Hugnot *et al.* [*Nucleic Acids Res.* **19**, 3159 (1991)], locus O (FRAXAC2) was as described [R. I. Richards *et al.*, *J. Med. Genet.* **28**, 818 (1991)], locus Q (GABRA3) was described by A. A. Hicks *et al.* [*Nucleic Acids Res.* **19**, 4016 (1991)], locus R (DXS52) was described by B. Richards *et al.* (*ibid.*, p. 1944), locus S (G6PD) was described by B. Kurdi-Haidar *et al.* [*Am. J. Hum. Genet.* **47**, 1013 (1990)], and locus T (F8C) was described by V. L. Surin *et al.* [*Nucleic Acids Res.* **18**, 3432 (1990)] The $(CCG)_n$-repeat at locus N (FMR) was assayed with the use of primers 203/213 described by E. J. Kremer *et al.* [*Science* **252**, 1711 (1991)] under PCR conditions described by S. Yu *et al.* [*Am. J. Hum. Genet.* **50**, 968 (1992)].

31. The data were analyzed by the likelihood ratio method of N. Risch [*Am. J. Hum. Genet.* **46**, 229 (1990)] as modified for X-linked markers in male sib-pairs. Let z_1 = the probability that a pair of brothers share a marker allele by-descent (z_1 is an unknown parameter that is estimated from the data), $\Gamma(z_1)$ = the likelihood of the observed data at z_1, $\Gamma(1/2)$ = the likelihood of the observed data under the null hypothesis of $z_1 = 1/2$, and $L(z_1) = \Gamma(z_1)/\Gamma(1/2)$. Then for the complete data set $2\ln L(z_1) =$

$$2\sum_{j=1}^{N} \ln[L_j(z_1)] =$$

$$2\left(\sum^{G} \ln[(1 - z_1)/(1/2)] + \sum^{D} \ln[(z_1)/(1/2)] + \sum_{j=1}^{S} \ln\{f_j(z_1 + f_j - f_j z_1)/[f_j(1 + f_j)/2]\}\right) =$$

$$2\left(G\ln(2 - 2z_1) + D\ln(2z_1) + \sum_{j=1}^{S} \ln[2z_1 + 2f_j - 2f_j z_1)/(1 + f_j)]\right)$$

where G is the number of discordant pairs, D is the number of concordant-by-descent pairs, S is the number of concordant-by-state pairs, f_j = the population frequency of the allele shared by the jth pair of concordant-by-state brothers, and $N = G + D + S$ = the number of informative pairs. This function was evaluated as a function of z_1 (between 0.5 and 1) to find the maximal value of $2\ln L(z_1)$. The one-sided significance P was estimated by considering $2\ln L(z_1)$ to be distributed as a chi-square at 1 degree of freedom.

32. We thank all of the participants, without whose cooperation and interest this research would not have been possible; W. Gahl, L. Charnas, S. Schlesinger, and the staff at the Whitman-Walker, NIAID, and NIH Interinstitute Genetics clinics for their assistance; D. Freije and H. Donis-Keller for communicating results prior to publication; W. McBride, C. Amos, J. Eldridge, and the staff of the Biomedical Supercomputer Center for technical and statistical advice; and E. Gershon, L. Goldin, J. Nathans, W. Gahl, D. Goldman, L. Charnas, E. Lander, M. Boehnke, F. Collins, members of the Laboratory of Biochemistry, and the reviewers for comments and suggestions on the manuscript.

2 April 1993; accepted 17 June 1993

APPENDIX B

INTERVIEW QUESTIONNAIRE

Your answers to the following questions are strictly confidential. If you do not wish to respond to a question, just tell me and we'll skip it.

PART A

1. What is your age?
2. How tall are you?
3. What is your weight?
4. What is your ethnic background?
   White (non-Hispanic) African-American/Black
   Asian/Pacific Islander Native-American/Alaskan
   Hispanic/Latino Other
5. How far have you been through school?
   grade school high school
   college graduate school

6. What is your occupation?
7. During the past year, was your income
   less than $15,000
   from $15,000 to $100,000
   more than $100,000
8. Which hand do you usually use for the following activities?

	LEFT	EITHER	RIGHT
write a letter	L	E	R
throw a ball	L	E	R
hold a racket	L	E	R
at the top of a push broom	L	E	R
on the lid of a jar as you unscrew it	L	E	R

9. Are there any activities you do with your other hand?
10. Were you ever forced to switch hands?

PART B

1. Have you ever seen a medical geneticist for any reason?
2. Are there any medical conditions that run in your family; e.g., cystic fibrosis, muscular dystrophy, etc.?
3. Is your red/green color vision OK?
4. Have you ever been tested for HIV?
5. What is your status?
6. If positive . . .
   When did you first test positive?
   What is your current immune-system status?
   Have you had Kaposi's sarcoma or lymphoma?
   Have you had any opportunistic infections?
   Have you had any other medical problems?
   Do you take any drugs or medications for HIV?

PART C

The following questions refer to your behavior as a child, 6 years old to 12 years old (K through 6th grade).

1. As a child, did you prefer to play with
   boys
   girls
   both about equally
   not to play with other children
2. As a child, did you like sports
   very much
   somewhat
   not at all
3. What sports did you prefer?
4. As a child, did you
   prefer games like cops and robbers, soldiers, war, etc.
   prefer games like dolls, cooking, sewing, etc.
   prefer both equally
5. What were your favorite play activities?
6. As a child, did you get in physical fights
   often
   occasionally
   rarely

7. As a child, were you as "boyish" or "masculine" as the other boys your age?
   more masculine than other boys
   about as masculine as other boys
   less masculine than other boys
   don't remember

8. As a child, were you called or considered a "sissy" more than other boys your age?
   no
   yes
   don't remember

9. Did you ever think you would prefer to be a girl rather than a boy?

Appendix B

PART D

1. At what age did you first have thoughts toward another person that you can now identify as at least partially sexual; e.g., a "crush" or "puppy love"?

2. Who were they toward?

3. Did you ever have such feelings toward a [person of the gender opposite to that of your first attraction]?

4. At what age did you first have any sexual contact that involved genitals?

5. What was the gender of your first sexual contact?
   female
   male
   mixed group
   never had sex, or don't remember

6. Did you have any sexual contacts with a [person of the gender opposite to that of your first sexual contact], and if so when?

7. As a child, were you ever approached by an older person for sex?

8. When did you reach puberty?

9. Was that about the same time as other boys?

10. When did you first ejaculate, and how did it occur?

11. After you reached puberty, if you had sexual thoughts, for example when masturbating, were they about
    females
    males
    both
    nonspecific, or no sexual thoughts

12. Did you date in high school? Who and how often?

13. Did you have girlfriends or boyfriends in high school?

14. Did you have sex in high school? If yes, with whom?
    females
    males
    both (% of each)

15. During high school, how did you regard your sexuality?

16. At what age did you acknowledge *to yourself* your sexual orientation?

17. Is there a reason that age was significant?

18. At what age did you start to acknowledge your sexual orientation to others?

19. What was coming out like for you?

20. Since that time, have there been any changes in your sexual orientation?

21. During your entire life, how many different sexual partners have you had?

    females
    males

22. Have you ever been sexually abused?

PART E

The following questions are about your current sex life, over the last year.

1. How often, on average, do you have sex with another person?
    once or more a day
    once or more a week
    once or more a month
    less than once a month

2. How many different sexual partners have you had over the last year?
    females
    males

3. Have your sexual contacts been in the context of
    a monogamous relationship
    a nonmonogamous relationship
    dating/casual romance
    anonymous sex

4. Do you or have you ever had a regular sexual partner, lover, or spouse?

5. How many and of what duration?
    females
    males

PART F

Now I'd like to ask some questions with numerical answers. [Give the participant the Kinsey scales.] This is a scale that goes from 0 to 6:

> **"0"** stands for someone who is exclusively heterosexual.
>
> **"1"** means someone who is predominantly heterosexual but every once in a while is interested in other men.
>
> **"2"** is for someone who identifies as heterosexual, but is attracted to or active with men more than just occasionally.
>
> **"3"** is fully bisexual, meaning equally interested in men and women.
>
> **"4"** is someone who is gay but is attracted to or active with women more than just occasionally.
>
> **"5"** is predominantly interested in or active with men.
>
> **"6"** is exclusively gay.

1. Overall, how would you identify yourself on this scale?
2. In the past, how would you have identified yourself?
3. How would you like to identify yourself in the future?

Here is another scale on which "0" stands for exclusively women and "6" stands for exclusively men.

4. Suppose you are walking down the street or at a party. You notice various people, some women and some men, and you think to yourself, "That one might be interesting to go to bed with." What mixture of men versus women would you be attracted to?
5. When you think about having sex with someone, for example while masturbating, what mixture of women versus men do you fantasize about?
6. What mixture of men and women do you actually have sex with?
7. In the past, what mixture of women and men did you have sex with?
8. In the future, what mixture of men and women would you like to have sex with?

PART G

Now I'd like to ask about specific sexual activities. Please remember that all the questions are optional. [Ask about before and after HIV, as appropriate.]

1. Do you ever have oral-genital sex? Compared to other sexual activities, is it something you do
   frequently
   occasionally
   rarely
   Do you more often
   do your partner
   get done
   mutual

2. Do you ever have penile-vaginal intercourse? Compared to other sexual activities, is it something you do
   frequently
   occasionally
   rarely

3. Do you ever have penile-anal intercourse? Compared to other sexual activities, is it something you do
   frequently
   occasionally
   rarely
   Do you more often
   insert into your partner
   have your partner insert into you
   mutual

4. Do you ever have oral-anal sex? Compared to other sexual activities, is it something you do
   frequently
   occasionally
   rarely
   Do you more often
   do your partner
   get done
   mutual

5. Is there a particular sexual activity that you would most *prefer* to do—assuming there were no AIDS or other STDs and that your partner was willing?

PART H

1. Do you drink coffee?
   no
   yes
   used to but quit
2. Do you smoke tobacco?
   no
   yes
   used to but quit
3. Do you drink alcohol?
4. If no . . . Is there a reason you don't drink?
5. If yes . . . How much do you drink? (A "drink" is a 12-ounce can of beer, an 8-ounce glass of wine, or 2 ounces of hard liquor.)
   per day
   per week
   per month
6. Do you feel you are a normal drinker?
7. Have you ever awakened the morning after some drinking the night before and not been able to remember everything that happened the night before?
8. Have you ever had trouble at work or been fired because of drinking?
9. Have you ever had fights or arguments because of drinking?
10. Do you ever find that you plan to drink a certain amount but end up drinking more or for a longer time?
11. Have you ever had a DUI or been arrested because of drinking?
12. Have you ever attended a meeting of Alcoholics Anonymous?
13. Have you ever sought help or been in a hospital because of drinking?
14. Do you take any psychoactive drugs?
   prescription (specify)
   pot
   coke (form)
   speed
   downers

15. Have you ever had a problem with drug use?

16. Have you ever seen a psychiatrist, psychologist, or therapist for any reason? Please explain.

17. Were you ever given a specific diagnosis, perhaps for insurance purposes?

18. Were you prescribed any drugs?

19. Have you ever had a period of six months or more when you felt sad or depressed more often than not?

20. Have you ever thought about suicide?

21. Have you ever attempted suicide?

22. Have you ever been admitted to the psychiatric ward of a hospital for any reason?

PART I

Now I'd like to ask some questions about your family. What we'll try to do is build a family tree, but instead of writing down people's names and birth dates, I'd like to find out what you know about their sexual orientation. You may know this for sure, you may be pretty certain but not definite, or you might not have any idea—so I'll also ask you how you got your information. Of course, I won't tell any of your relatives what you said about them or vice versa.

1. How many brothers do you have?

2. Do you think any of them might be gay or bisexual?

3. Why do you think that?

4. [Continue collecting information for each relative and record on pedigree chart]:

    brothers
    sisters
    nephews (enumerate by sibling)
    nieces (enumerate by sibling)
    father
    mother
    paternal uncles
    paternal aunts
    paternal cousins through uncle (enumerate by uncle)

paternal cousins through aunt (enumerate by aunt)
paternal grandfather
paternal grandmother
maternal uncles
maternal aunts
maternal cousins through uncle (enumerate by uncle)
maternal cousins through aunt (enumerate by aunt)
maternal grandfather
maternal grandmother

5. Do you know of any additional family members who are more distantly related that you think might be gay, lesbian, bisexual, or questionable?

6. Why do you think that?

7. Are any of your nuclear family members left-handed?

8. Do you think any of your relatives have a problem with alcohol or drugs?

9. What is the ethnic background of your parents?

PART J

1. How would you characterize your relationship with your mother, growing up?

2. How would you characterize your relationship with your father, growing up?

3. Were you closer to your mother or your father, growing up?

4. How would you characterize your relationship with your brothers, growing up?

5. How would you characterize your relationship with your sisters, growing up?

6. Did your mother spend more time with you or your siblings (specify which siblings)?

7. Did your father spend more time with you or your siblings (specify which siblings)?

8. Is there anything else you'd like to tell me about your family or growing up?

NOTES

CHAPTER THREE: WHO'S GAY?

Page

53 *who was in between:* Taken from the title of the book *Gay, Straight, and In-between,* by J. Money (New York: Oxford University Press, 1988).

CHAPTER FIVE: A MOTHER'S LEGACY

Page

104 *it made little difference which background rate I used:* A chi-square test was used to calculate the statistical significance of the differences between the rates of homosexuality observed in the relatives of the gay male subjects and the estimated population rates of 2.0% or 2.6%. An exact test was used to compare the rates found in the relatives of the gay men to the rates in the corresponding group of lesbian relatives; e.g., brothers of gay men versus brothers of lesbians, maternal uncles of gay men versus maternal uncles of lesbians, etc. Both tests generate a P value, which is a measure of the probability that the two samples are drawn from equivalent populations. For example, a P value of 0.05 indicates a 95% probability

that the difference between the two groups is real, while a P value of 0.001 indicates a 99.9% confidence level. The calculated P values are shown below.

		Significance Level		
RELATIVES	**% GAY**	**P VS. 2.0% BCKG.**	**P VS. 2.6 % BCKG.**	**P VS. LESB. RELS.**
Fathers	0.0	>.05	>.05	>.05
Brothers	13.5	<.001***	<.001***	.010**
Maternal uncles	7.3	.002**	.013*	.014*
Paternal uncles	1.7	>.05	>.05	>.05
Maternal cousins thru aunt	7.7	.010**	.036*	.043*
Maternal cousins thru uncle	3.9	>.05	>.05	>.05
Paternal cousins thru aunt	3.6	>.05	>.05	>.05
Paternal cousins thru uncle	5.4	>.05	>.05	>.05

* $P \leq .05$
** $P \leq .01$
*** $P \leq .001$

CHAPTER SIX: LOOKING FOR LINKAGE

Page

109 *an enrichment of more than tenfold:* Consider an X-linked gene with two alleles, X1 and X2. Let the trait-associated allele X1 have a population frequency of p and a penetrance of u. Let the non-trait-associated allele X2 have a frequency of q (where $p + q = 1$) and a penetrance of v (where $v < u$). The population frequency of the trait, k, is then given by $k = pu + qv$. The frequency pT of the X1 allele among all men who express the trait is $pT = pu/k$. By contrast, the frequency pS of the X1 allele in men who have a brother who also expresses the trait is given by pS =

$$\{p^2u^2 + 2pq[0.25(u^2 + 2uv)]\}/\{p^2u^2 + q^2v^2 + 2pq[0.25(u^2 + v^2 + 2uv)]\}$$

This equation can be used to calculate the enrichment, pS/pT, for the trait-associated allele in shared-trait sib-pairs. For the example in which the X-linked gene accounts for 50% of gays, we let $p = 0.012$, $u = 0.85$, and $v = 0.01$ giving $k = 0.02$, $pT = 0.5$, $pS = 0.97$. Hence the enrichment for the X1 allele in sib-pair families is $pS/pT = 0.97/0.5 = 1.94$. For the example in which the gene accounts for only 5% of gays, we let $p = 0.0012$, $u = 0.85$, $v = 0.019$ giving $k = 0.02$, $pT = 0.05$, $pS = 0.55$. In this case the enrichment for the X1 allele in sib-pairs is $pS/pT = .55/.05 = 11$.

118 *evidence for linkage has disappeared:* For the color blindness pedigree shown in fig. 4, the maximum LOD score is 3.4 at a recombination fraction of 0, which indicates tight linkage between color blindness and the marker. However, suppose that one of the uncles has a second gene that restores his color vision, and that one of his nephews is mistakenly assessed as color-sighted (these two individuals are indicated in the pedigree by black squares with white dots). In that case, the LOD score at a recombination fraction of 0 would be negative infinity, which would be interpreted as evidence *against* tight linkage between the marker and color blindness. Furthermore, the maximum LOD score (at a recombination fraction of 0.19) would be 0.78, which is not sufficient to show *any* linkage.

120 *to prove the same linkage:* Let α be the fraction of a population of shared-trait male sib-pairs that is "linked to" an X-chromosome marker; that is, α is the fraction of this population that displays excess coinheritance of the marker. Let p be the population frequency of the trait-associated allele and let t be the recombination fraction between the trait-gene and the marker. Then the ratio R of concordant-by-state to discordant sib-pairs is

$$R = \{(1-\alpha)/2 + \alpha(P/2 + (1-P)T\}/\{(1-\alpha)/2 + \alpha(P/2 + (1-P)(1-T)\}$$

where $P = 2p^2/(p + p^2)$ and $T = 1 - 2t + 2t^2$. When $\alpha = 0.67$ and $p = 0.02$, $t = 0$ gives $R = 4.6$ whereas $t = 0.1$ gives $R = 2.4$. This decrease in R increases the sample size needed to obtain a LOD score of 3.0 from $N = 30$ when $t = 0$ to $N = 81$ when $t = 0.1$. At a recombination fraction of $t = 0.2$, a sample size of greater than 200 informative pairs would be required to obtain a LOD score of 3.0.

CHAPTER EIGHT: GOING PUBLIC

Page

141 *"linked to" the Xq28 region:* Let α be the fraction of a population of shared-trait male sib-pairs that is "linked to" an X-chromosome marker.

Ignoring the effects of recombination and of maternal homozygosity for the trait-associated allele, both of which will decrease the estimate of α, the fraction of sib-pairs that are concordant for the marker, C, is given by $C = 1\alpha + 0.5(1-\alpha)$. (This is equivalent to the expression in the previous note when $p = t = 0$.) This is because all of the linked pairs and half of the unlinked pairs will be concordant. Therefore $\alpha = 2C-1$. In our study $C = 33/40$, giving $\alpha = 0.65 = 65\%$ linked.

146 *nurture rather than nature:* J. Maddox, "Wilful Misunderstanding of Genetics" [editorial], *Nature* 365 (1993):281.

148 *"homosexuality in the general population":* M.-C. King, "Sexual Orientation and the X" [news], *Nature* 364 (1993):288–89.

CHAPTER TEN: PSYCHOLOGICAL MECHANISMS: SISSIES, FREUD, AND SEX ACTS

Page

182 *how much is environmental:* Looking across a single row of this matrix gives the "norm of reaction" for one particular genotype; that is, a list of phenotypes that will result when an individual of this genotype develops in different alternative environments. Although R. C. Lewontin et al., in *Not in Our Genes* (New York: Pantheon Books, 1984), 268, state that the norm of reaction represents "The fundamental concept for understanding the relationship between gene, environment, and organism . . . ," it actually is useless in understanding the contribution of genetic variability to phenotypic variability. This can be calculated only by looking at the entire matrix, including all possible genotypes.

CHAPTER ELEVEN: EVOLUTION

Page

185 *They only think about themselves:* See R. Dawkins, *The Selfish Gene* (Oxford: Oxford University Press, 1976), and *The Extended Phenotype* (San Francisco: W. H. Freeman, 1981).

188 *evolutionary process of sexual selection:* W. B. Rice, in "Sex Chromosomes and the Evolution of Sexual Dimorphism," *Evolution* 38 (1984):735–42, has compared the evolution of autosomal and X-linked sexually antagonistic genes. Consider an initially rare allele of a gene that reduces the reproduction of men by 2-fold. If the gene is autosomal dominant, it must increase female reproduction by more than 2-fold to increase. However, if the gene is X-linked dominant, it will increase so

long as it benefits female reproduction by more than 1.33-fold; for example, if it benefits women by 1.35-fold, it will reach an equilibrium level of 0.036, and 3.6% of men will be gay. For genes that have a smaller initial effect on male reproduction, the benefit to women can be as little as one half the cost to men. Interestingly, genes that lie in a pseudoautosomal region of X-Y homology have effects that are intermediate between the autosomal and sex-linked cases. For example, I used simulations to show that a dominant pseudoautosomal gene that undergoes X-Y recombination in 2% of male meioses and that reduces male reproduction by 2-fold and increases female reproduction by 1.35-fold will reach an equilibrium level of 0.014 on X chromosomes and 0.0003 on Y chromosomes to give 1.7% gay men. A recessive pseudoautosomal gene that is tightly linked to sex can reach significant equilibrium levels even when male reproduction is reduced to zero and female reproduction is increased by only 10% (unpublished calculations).
190 *social behavior of some insects:* Kin selection is based on the broader concept of inclusive fitness and has been most convincingly demonstrated for social insects in W. D. Hamilton, "The Genetical Evolution of Social Behavior," *Journal of Theoretical Biology* 7 (1964):1–52. The possible role of kin selection in the evolution of human sexual orientation has been discussed by Trivers in "Parent-Offspring Conflict," *American Zoologist* 14 (1974):249–64; E. O. Wilson in *On Human Nature* (Cambridge: Harvard University Press, 1978); and J. D. Weinrich in "Nonreproduction, Homosexuality, Transsexualism, and Intelligence: I. A Systematic Literature Search," *Journal of Homosexuality* 3 (1978):275–89, and in "A New Sociobiological Theory of Homosexuality Applicable to Societies with Universal Marriage," *Ethology and Sociobiology* 8 (1987):37–47.

CHAPTER TWELVE: BEYOND SEX

Page
196 *blood relative with a problem:* Interestingly, R. C. Pillard, In "Sexual Orientation and Mental Disorder," *Psychiatric Annals* 18 (1988):52–56, found an increased rate of alcoholism in the relatives of gay men.
204 *parent-offspring pairs and adoptees:* An excellent review of this research is found in Plomin et al., *Behaviorial Genetics* (New York: W. H. Freeman, 1990).

CHAPTER THIRTEEN: BEYOND THE LAB: IMPLICATIONS OF A "GAY GENE"

Page
224 *predicted by a blood test:* Although the *overall* accuracy of such a test would never be greater than 50 percent, it could be considerably higher or lower in particular individuals.

SOURCES AND FURTHER READING

SEXUAL ORIENTATION: GENERAL

Bell, A. P., and M. S. Weinberg. 1978. *Homosexualities: A study of diversity among men and women.* New York: Simon and Schuster.

Bell, A. P., M. S. Weinberg, and S. K. Hammersmith. 1981. *Sexual preference: Its development in men and women.* Bloomington, IN: Indiana University Press.

Isay, R. A. 1989. *Being homosexual: Gay men and their development.* New York: Farrar, Straus, and Giroux.

Money, J. 1988. *Gay, straight, and in-between: The sexology of erotic orientation.* New York: Oxford University Press.

Tripp, C. A. 1975. *The homosexual matrix.* New York: Signet.

Weinrich, J. D. 1987. *Sexual landscapes.* New York: Scribners.

SEXUAL ORIENTATION: GENETICS

Bailey, J. M., and D. S. Benishay. 1993. Familial aggregation of female sexual orientation. *American Journal of Psychiatry* 150:272–77.

Bailey, J. M., and R. C. Pillard. 1991. A genetic study of male sexual orientation. *Archives of General Psychiatry* 48:1089–96.

———. 1993. A genetic study of male sexual orientation [letter]. *Archives of General Psychiatry* 50:240–41.

Bailey, J. M., R. C. Pillard, M. C. Neale, and Y. Agyei. 1993. Heritable factors influence sexual orientation in women. *Archives of General Psychiatry* 50:217–23.

Henry, G. W. 1948. *Sex variants: A study of homosexual patterns.* New York: Paul B. Hoeber.

Macke, J. P., N. Hu, S. Hu, M. Bailey, V. L. King, T. Brown, D. Hamer, and J. Nathans. 1993. Sequence variation in the androgen receptor gene is not a common determinant of male sexual orientation. *American Journal of Human Genetics* 53:884–92.

Pillard, R. C., and J. D. Weinrich. 1986. Evidence of familial nature of male homosexuality. *Archives of General Psychiatry* 43:808–12.

Whitam, F. L., M. Diamond, and J. Martin. 1993. Homosexual orientation in twins: A report on 61 pairs and 3 triplet sets. *Archives of Sexual Behavior* 22:187–206.

SEXUAL ORIENTATION: MEASUREMENT, PSYCHOLOGY, AND CORRELATES

Bailey, J. M., and K. J. Zucker. 1994. Childhood sex-typed behavior and sexual orientation: A conceptual analysis and quantitative review. *Developmental Psychology.* In press.

Becker, J. T., S. M. Bass, M. A. Dew, L. Kingsley, O. A. Selnes, and K. Sheridan. 1992. Hand preference, immune system disorder and cognitive function among gay/bisexual men: The multicenter AIDS cohort study (MACS). *Neuropsychologia* 30:229–35.

Cattell, R. B., and J. H. Morony. 1962. The use of the 16 PF in distinguishing homosexuals, normals, and general criminals. *Journal of Consulting Psychology* 26:531–40.

Duckitt, J. H., and L. Du Toit. 1989. Personality profiles of homosexual men and women. *The Journal of Psychology* 123:497–505.

Evans, R. B. 1970. Sixteen personality factor questionnaire scores of homosexual men. *Journal of Consulting and Clinical Psychology* 34:212–15.

Freud, S. [1905] 1953. "Three essays on the theory of sexuality." In *Standard edition of the complete psychological works of Sigmund Freud.* Vol. 7, 125–243. London: Hogarth Press.

Freund, K. W. 1967. Diagnosing homo- or heterosexuality and erotic age-preference by means of a psychophysiological test. *Behavior Research and Therapy* 5:209–28.

Gonsiorek, J. C., and J. D. Weinrich. 1991. "The definition and scope of sexual orientation." In *Homosexuality: Research implications for public policy,* ed. J. C. Gonsiorek and J. D. Weinrich. Newbury Park, CA: Sage.

Green, R. 1987. *The "sissy boy syndrome" and the development of homosexuality.* New Haven, CT: Yale University Press.

Kinsey, A. C., W. B. Pomeroy, and C. E. Martin. 1948. *Sexual behavior in the human male.* Philadelphia: W. B. Saunders.

Kourany, R. F. 1987. Suicide among homosexual adolescents. *Journal of Homosexuality* 13 (4): 111–17.

Lindesay, J. 1987. Laterality shift in homosexual men. *Neuropsychologia* 25:965–69.

Marchant-Haycox, S. E., I. C. McManus, and G. D. Wilson. 1991. Left-handedness, homosexuality; HIV infection and AIDS. *Cortex* 27:49–56.

McCormick, C. M., S. F. Witelson, and E. Kingstone. 1990. Left-handedness in homosexual men and women: Neuroendocrine implications. *Psychneuroendocrinology* 15:69–76.

Pillard, R. C. 1988. Sexual orientation and mental disorder. *Psychiatric Annals* 18:52–56.

Rofes, E. E. 1983. *I thought people like that killed themselves.* San Francisco: Grey Fox Press.

Socarides, C. W. 1978. *The overt homosexual.* New York: Grune and Stratton.

Whitam, F. L., and R. M. Mathy. 1986. *Male homosexuality in four societies: Brazil, Guatemala, Philippines, and the United States.* New York: Praeger.

SEX, GENES, HORMONES, AND THE BRAIN

Allen, L. S., and R. A. Gorski. 1991. Sexual dimorphism of the anterior commissure and massa intermedia of the human brain. *Journal of Comparative Neurology* 312:97–104.

———. 1992. Sexual orientation and the size of the anterior commissure in the human brain. *Proceedings of the National Academy of Sciences, USA* 89:7199–202.

Allen, L. S., M. Hines, J. E. Shryne, and R. A. Gorski. 1989. Two sexually dimorphic cell groups in the human brain. *Journal of Neuroscience* 9:497–506.

Dohler, K.-D., A. Coquelin, F. Davis, M. Hines, J. E. Shryne, and R. A. Gorski. 1984. Pre- and postnatal influence of testosterone propionate and diethylstilbesterol on differentiation of the sexually dimorphic nucleus of the preoptic area in male and female rats. *Brain Research* 302:291–95.

Gorski, R. A., R. E. Harlan, C. D. Jacobsen, J. E. Shryne, and A. M. Southam. 1980. Evidence for a morphological sex difference within the medical preoptic area of the rat brain. *Journal of Comparative Neurology* 193:529–39.

Koopman, P., J. Gubbay, N. Vivian, P. Goodfellow, and R. Lovell-Badge. 1991. Male development of chromosomally female mice transgenic for Sry. *Nature* 351:117–21.

LeVay, S. A difference in hypothalamic structure between heterosexual and homosexual men. *Science* 253:1034–37.

———. 1993. *The sexual brain.* Cambridge, MA: MIT Press.

Meyer-Bahlburg, H. F. L. 1982. Psychoendocrine research on sexual orientation: Current status and future options. *Progress in Brain Research* 61:375–98.

Money, J., M. Schwartz, and V. G. Lewis. 1984. Adult erotosexual status and fetal hormonal masculinization and demasculization: 46, XX congenital virilizing adrenal hyperplasia and 46, XX androgen-insensitivity syndrome compared. *Psychoneuroendocrinology* 9:405–14.

Sinclair, A. H., P. Berta, M. S. Palmer, J. R. Hawkins, B. L. Griffiths, M. J. Smith, J. W. Foster, A. M. Frischauf, R. Lovell-Badge, and P. N. Goodfellow. 1990. A gene from the human sex-determining region encodes a protein with homology to a conserved DNA-binding motif. *Nature* 346:240–42.

Swaab, D. F., and M. A. Hofman. 1990. An enlarged suprachiasmatic nucleus in homosexual men. *Brain Research* 537:141–48.

HUMAN GENETICS: MARKERS, DISEASES, AND THE GENOME

Freije, D., C. Helms, M. S. Watson, and H. Donis-Keller. 1992. Identification of a second pseudoautosomal region near the Xq and Yq telomeres. *Science* 258:1784–87.

Fu, Y.-H., D. P. A. Kuhl, A. Pizzuti, M. Peiretti, J. S. Sutcliffe, S. Richards, A. J. M. H. Verkerk, J. J. A. Holden, R. G. Fenwich, Jr., S. T. Warren, B. A.

Oostra, D. L. Nelson, and C. T. Caskey. 1991. Variation of the CGG repeat at the fragile X site results in genetic instability: Resolution of the Sherman paradox. *Cell* 67:1047–58.

Gelehrter, T. D., and F. S. Collins. 1990. *Principles of medical genetics.* Baltimore: Williams and Wilkins.

Hall, J. M., M. K. Lee, B. Newman, J. E. Morrow, L. A. Anderson, B. Huey, and M.-C. King. 1990. Linkage of early-onset familial breast cancer to chromosome 17q21. *Science* 250:1684–89.

Kremer, E. J., M. Pritchard, M. Lynch, S. Yu, K. Holman, E. Baker, S. T. Warren, D. Schlessinger, G. R. Sutherland, and R. I. Richards. 1991. Mapping of DNA instability at the fragile X to the trinucleotide repeat sequence p(CCG)n. *Science* 252:1711–14.

La Spada, A. R., E. M. Wilson, D. B. Lubahn, A. E. Harding, and K. H. Fischbeck. 1991. Androgen receptor gene mutations in X-linked spinal and bulbar muscular atrophy. *Nature* 352:77–79.

Olson, M. V. 1993. The human genome project. *Proceedings of the National Academy of Sciences, USA* 90:4338–44.

Singer, M., and P. Berg. 1991. *Genes and genomes.* Mill Valley, CA: University Science Books.

Singer, S. 1985. *Human genetics.* New York: W. H. Freeman.

Verkerk, A. J. M. H., M. Pieretti, J. S. Sutcliffe, Y.-H. Fu, D. P. A. Kuhl, A. Pizzuti, O. Reiner et al. 1991. Identification of a gene (FMR-1) containing a CGG repeat coincident with a breakpoint cluster exhibiting length variation in fragile X syndrome. *Cell* 65:905–14.

Watson, J. D. 1990. The human genome project: Past, present, and future. *Science* 248:44–49.

HUMAN GENETICS: LINKAGE ANALYSIS

Botstein, D., R. L. White, M. Skolnick, and R. W. Davis. 1980. Construction of a genetic linkage map in man using restriction fragment length polymorphisms. *American Journal of Human Genetics* 32:314–31.

Lander, E. S. 1988. "Mapping complex genetic traits in humans." In *Genome analysis,* ed. K. E. Davis, 171–89. New York: IRL Press.

Lander, E. S., and D. Botstein. 1986. Strategies for studying heterogeneous traits in humans using a linkage map of restriction fragment length polymorphisms. *Proceedings of the National Academy of Sciences, USA* 83:7353–57.

Ott, J. *Analysis of human genetic linkage.* 1991. Baltimore: Johns Hopkins University Press.

HUMAN GENETICS: BEHAVIOR AND PSYCHIATRY

Baron, M., N. Freimer, N. Risch, B. Lerer, J. Alexander, R. Straub, S. Asokan, K. Das, A. Peterson, J. Amos, J. Endicott, J. Ott, and T. Gilliam. 1993. Diminished support for linkage between manic-depressive illness and X-chromosome markers in three Israeli pedigrees. *Nature Genetics* 3:49–55.

Baron, M., N. Risch, R. Hamburger, B. Mandel, S. Kushner, M. Newman, D. Drumer, and R. Belmaker. 1987. Genetic linkage between X-chromosome markers and bipolar affective illness. *Nature* 326:289–92.

Bouchard, T. J., and M. McGue. 1981. Familial studies of intelligence: A review. *Science* 212:1055–59.

Brunner, H. G., M. Nelson, X. O. Breakefield, H. H. Ropers, and B. A. van Oost. 1993. Abnormal behavior associated with a point mutation in the structural gene for monoamine oxidase A. *Science* 262:578–80.

Corder, E. H., A. M. Saunders, W. J. Strittmatter, D. E. Schmechel et al. 1993. Gene dose of apolipoprotein E type 4 allele and the risk of Alzheimer's disease in late-onset families. *Science* 261:921–23.

Egeland, J., D. Gerhard, D. Pauls, J. Sussex et al. 1987. Bipolar affective disorders linked to DNA markers on chromosome 11. *Nature* 325:783–87.

Henderson, N. D. 1982. Human behavior genetics. *Annual Review of Psychology* 33:403–40.

Kagan, J., and N. Snidman. 1991. Temperamental factors in human development. *American Psychology* 46:856–62.

Kelsoe, J. R., E. Ginns, J. Egeland, D. Gerhard et al. 1989. Re-evaluation of the linkage relationship between chromosome 11p loci and the gene for bipolar affective disorder in the Old Order Amish. *Nature* 16:238–43.

Linnoila, M., M. Virkkunen, T. George, and D. Higley. 1993. Impulse control disorders. *International Clinical Psychopharmacology* 8 (suppl. 1):s53–56.

Loehlin, J. C., L. Willerman, and J. M. Horn. 1988. Human behavior genetics. *Annual Review of Psychology* 39:101–33.

McManus, I. C., and M. P. Bryden. 1992. "The genetics of handedness, cerebral dominance and lateralization." In *Handbook of neuropsychology, Vol. 6, Sec. 10: Child neuropsychology,* ed. I. Rapin and S. G. Segalowitz, 115–44. Amsterdam: Elsevier.

Mullan, M., and F. Crawford. 1993. Genetic and molecular advances in Alzheimer's disease. *Trends in Neuroscience* 16:398–403.

Nielsen, D. A., D. Goldman, M. Virkkunen, R. Tokola et al. 1994. Suicidality and 5-hydroxyindoleacetic acid concentration associated with a tryptophan hydroxylase polymorphism. *Archives of General Psychiatry* 51:34–38.

Pericak-Vance, M. A., J. L. Bebout, P. C. Gaskell, L. H. Yamaoka et al. 1991. Linkage studies in familial Alzheimer's disease: Evidence for chromosome 19 linkage. *American Journal of Human Genetics* 48:1034–50.

Plomin, R. 1990. The role of inheritance in behavior. *Science* 248:183–88.

Plomin, R., J. C. DeFries, and G. E. McClearn. 1990. *Behavioral Genetics.* New York: W. H. Freeman.

Strittmatter, W. J., A. M. Saunders, D. Schmechel., M. Pericak-Vance et al. 1993. Apolipoprotein E: High-avidity binding to beta-amyloid and increased frequency of type 4 allele in late-onset familial Alzheimer's disease. *Proceedings of the National Academy of Sciences, USA* 90:1977–81.

Tellegan, A., D. T. Lykken, T. J. Bouchard, K. Wilcox, N. Segal, and S. Rich. 1988. Personality similarity in twins reared apart and together. *Journal of Social and Personality Psychology* 54:1031–39.

ALCOHOLISM

Blum, K., E. P. Noble, P. J. Sheridan et al. 1990. Allelic association of human dopamine D2 receptor gene in alcoholism. *Journal of the American Medical Association* 263:2055–60.

Bolos, A. M., M. Dean, S. Lucas-Derse, M. Ramsburg, G. L. Brown, and D. Goldman. 1990. Population and pedigree studies reveal a lack of association between the dopamine D2 receptor gene and alcoholism. *Journal of the American Medical Association* 264:3156–60.

Gejman, P., A. Ram, J. Gelernter, E. Friedman, Q. Cao, D. Pickar, K. Blum, E. Noble, H. Kranzler, S. O'Malley, D. Hamer, F. Whitsit, P. Rao, L. DeLisi, M. Virkkunen, M. Linnoila, D. Goldman, and E. Gershon. 1994. No structural mutations in the dopamine D2 receptor gene in alcoholism or schizophrenia. *Journal of the American Medical Association* 271:204–8.

Gelernter, J., S. O'Malley, N. Risch et al. 1991. No association between an allele at the D2 dopamine receptor gene (DRD2) and alcoholism. *Journal of the American Medical Association* 266:1801–7.

McKirnan, D., and P. Peterson, 1989. Alcohol and drug use among homosexual men and women: Epidemiology and population characteristics. *Addictive Behaviors* 14:545–53.

HIV/AIDS

Gallo, R. 1991. *Virus Hunting.* New York: Basic Books.

Hentges, F., A. Hoffmann, F. Oliveira de Araujo, and R. Hemmer. 1992. Prolonged clinically asymptomatic evolution after HIV-1 infection is marked by the absence of complement C4 null alleles at the MHC. *Clinical and Experimental Immunology* 88:237–42.

Jeannet, M., R. Sztajzel, N. Carpentier, B. Hirschel, and J. M. Tiercy. 1989. HLA antigens are risk factors for development of AIDS. *Journal of Acquired Immune Deficiency Syndrome* 2:28–32.

Kuntz, B. M., and H. T. Bruster. 1989. Time-dependent variation of HLA antigen frequencies in HIV-1 infection (1983–1988). *Tissue Antigens* 34:164–69.

Louie, L. G., B. Newman, and M.-C. King. 1991. Influence of host genotype on progression to AIDS among HIV-infected men. *Journal of Acquired Immune Deficiency Syndrome* 4:814–18.

Mann, D. L., C. Murray, M. O'Donnell, W. A. Blattner, and J. J. Goedert. 1990. HLA antigen frequencies in HIV-1-related Kaposi's sarcoma. *Journal of Acquired Immune Deficiency Syndrome* 3 (suppl. 1):s51–55.

EVOLUTION

Darwin, C. (1871) 1981. *The descent of man, and selection in relation to sex.* Princeton, NJ: Princeton University Press.

Dawkins, R. 1976. *The selfish gene.* Oxford: Oxford University Press.

———. 1981. *The extended phenotype.* San Francisco: W. H. Freeman.

Hamilton, W. D. 1964. The genetical evolution of social behavior. *Journal of Theoretical Biology* 7:1–52.

Rice, W. R. 1984. Sex chromosomes and the evolution of sexual dimorphism. *Evolution* 38:735–42.

Trivers, R. L. 1974. Parent-offspring conflict. *American Zoologist* 14:249–64.

———. 1985. *Social evolution.* Menlo Park, CA: Benjamin/Cummings.

Weinrich, J. D. 1978. Nonreproduction, homosexuality, transsexualism, and intelligence: I. A systematic literature search. *Journal of Homosexuality* 3:275–89.

———. 1987. A new sociobiological theory of homosexuality applicable to societies with universal marriage. *Ethology and Sociobiology* 8:37–47.

Wilson, E. O. 1978. *On human nature.* Cambridge: Harvard University Press.

MISCELLANEOUS

Furst, P., S. Hu, R. Hackett, and D. Hamer. 1988. Copper activates metallothionein gene transcription by altering the conformation of a specific DNA-binding protein. *Cell* 55:705–17.

Hubbard, R., and E. Wald. 1993. *Exploding the gene myth.* Boston: Beacon Press.

Lewontin, R. C., S. Rose, and L. J. Kamin. 1984. *Not in our genes.* New York: Pantheon Books.

INDEX